A MANUAL OF HUMAN ANATOMY
VOLUME II

HEAD AND NECK

A MANUAL OF HUMAN ANATOMY

VOLUME II

HEAD AND NECK

By

J. T. AITKEN, M.D.
PROFESSOR OF ANATOMY AT
UNIVERSITY COLLEGE, LONDON

G. CAUSEY, M.B., F.R.C.S.
PROFESSOR EMERITUS,
UNIVERSITY OF LONDON

J. JOSEPH, M.D., D.Sc., F.R.C.O.G.
PROFESSOR OF ANATOMY AT
GUY'S HOSPITAL MEDICAL SCHOOL,
LONDON

J. Z. YOUNG, M.A., D.Sc., LL.D., F.R.S.
PROFESSOR EMERITUS
UNIVERSITY OF LONDON

THIRD EDITION

CHURCHILL LIVINGSTONE
EDINBURGH LONDON AND NEW YORK
1976

CHURCHILL LIVINGSTONE

Medical Division of Longman Group Limited

Distributed in the United States of America by
Longman Inc., 19 West 44th Street, New York,
N.Y. 10036 and by associated companies,
branches and representatives throughout the world.

First Edition   .   .   1957
Second Edition .   .   1964
  Reprinted   .   .   1968
  Reprinted   .   .   1972
Third Edition .   .   1976
  Reprinted   .   .   1978

ISBN 0 443 01241 5

British Library Cataloguing in Publication Data

A manual of human anatomy.—3rd ed.
  Vol. 2: Head and neck.
  1. Anatomy, Human
  I. Aitken, John Thomas
  611   QM23.2   74-33179

Printed in Great Britain by
Butler & Tanner Ltd, Frome and London

PREFACE TO THE THIRD EDITION

SINCE the publication of the Second Edition in 1964, there have been two reprints of the volumes. The continued confidence shown in this demand is most gratifying to the authors and justifies their original conception of the scope of a preclinical course in topographical anatomy.

In many departments radical changes are being made in the methods of teaching anatomy and in the content of the course. These experiments are to be welcomed. Usually the time available for careful dissection has been reduced and many students have to be content with a more rapid and less detailed approach to the body. The authors suggest that the dissection of some regions may be omitted, although the dissecting instructions are retained in the text.

In this Edition, the number of volumes has been reduced and the contents re-arranged. Volume I contains the thorax, abdomen and pelvis, Volume II the head and neck and Volume III the limbs. It is hoped that this arrangement will keep the total cost as low as possible and not inconvenience students or departments.

We are grateful to colleagues and students for helpful suggestions, and to Mrs J. Astafiev of University College, London, for the new drawings in the Second and Third Editions, and to the staff of Churchill Livingstone who continue to be most co-operative.

THE AUTHORS.

London,
1975

PREFACE TO THE FIRST EDITION

THE purpose of these Manuals is to give the student of human anatomy a method of dissecting the body and to act as a guide to the extent of the knowledge expected of him in the second medical examination. An attempt has been made to link together the structure and function of the different parts of the body, and the anatomy necessary for a future study of clinical medicine or an understanding of the development of the part is emphasised. Paragraphs indicating the functions of the parts under consideration have, where appropriate, been introduced after the practical instructions and topographical details. The study of each part can thus be undertaken with some knowledge of the functional implications of the anatomy and not as a mere exercise of memory. Much detail has been omitted.

A co-ordinated course is more easily organised if all the members of the class are dissecting the same part at the same time and the instructions in the Manuals are presented on this assumption. A most important part of the teaching is carried out by means of small classes on osteology and surface anatomy. For these classes we have found it useful to indicate what the students should know. The students do the work themselves and it is then checked by a demonstrator. Appropriate lists for such work are found at the end of each volume.

It has been found advantageous to begin with the dissection of the thorax. This results in an early acquaintance with the heart and lungs and with the peripheral and autonomic nervous systems, all of which receive attention in most introductory courses of physiology. From the thorax, the student proceeds to dissect the upper limb (vol. I), the head and neck (vol. II), the abdomen and pelvis (vol. III) and the lower limb (vol. IV). The descriptions and instructions in the Manuals assume that this order has been followed. Instructions for the use of the Manuals where a different order is employed are given opposite page 1. The gross and histological structure of the brain and spinal cord are described in vol. V.

Each part of the body is subdivided for convenience into smaller regions. In the limbs these regions centre around the joints and in other parts around the larger morphological or functional units. In each region, a short introduction is followed by dissecting instructions, including a description of many of the structures being dissected. There follow paragraphs on further details and relations of the structures, and their functions.

Summaries of the cutaneous nerve supply and of the lymphatic drainage of the part dissected are found towards the end of. each section of the Manual.

In the early stages of the planning and writing of these Manuals, Dr. W. A. Fell, now of Addenbrooke's Hospital, Cambridge, and Dr. D. H. L. Evans of University College, London, contributed to the work and much helpful criticism has been received from other colleagues and students.

The illustrations were produced by Miss E. R. Turlington and Miss J. de Vere, largely from specimens and drawings in the Anatomy Department at University College, London. As the main object of the pictures is to illustrate the text, all unnecessary complicating details have been omitted and the salient features emphasised by the use of colour.

Our thanks are also due to Miss A. Baxter and Miss M. Lynn for typing the final draft of the Manuals, and the staff of E. & S. Livingstone for the production and publication of the Manuals.

THE AUTHORS.

London,
February, 1957.

CONTENTS

ix

NOTE

THE order in which the different parts of the body are dissected varies. Many Departments prefer all their students to dissect the same part at the same time. By beginning with the thorax (Vol. I) the students are quickly introduced to the organs of respiration and circulation, and also to the spinal and autonomic parts of the nervous system. From the thorax, dissection can proceed to the abdomen and pelvis and later to the neck and head (Vol. II) and then to the limbs (Vol. III).

In some Departments, where different groups of students dissect the various parts of the body at the same time, some re-arrangement of the order is required. If dissection begins with the body on its back, dissection of the head and neck can begin at Volume II, Chapter 4. Later, with the dissectors of the upper limb, dissection proceeds to Volume III, Chapter 5 and then to Volume II, Chapter 2.

If dissection begins with the body on its face, work with the dissectors of the upper limb and begin with Volume III, Chapter 5 and proceed to Volume II, Chapter 2.

ORIENTATION

TO help in the description of a structure or a region certain terms are used and they have an agreed interpretation. The **anatomical position** is one in which the person stands upright, with the feet together, the eyes looking forward, and the arms straight along the sides of the body and the palms of the hands directed forwards. The front of the body is called the **anterior** surface and the back is called the **posterior** surface (see outside front cover drawing). The terms **ventral** and **dorsal** may be used for the front and back respectively. Higher structures are **superior** and lower structures are **inferior.** **Median** structures are found in the midline of the body and the terms **medial** (nearer to) and **lateral** (further from) are relative to the midline.

A **sagittal plane** passes vertically anteroposteriorly through the body and movements in this plane (see inside front cover drawing) are called **flexion** (forwards) or **extension** (backwards). A vertical plane at right angles to the sagittal is called a **coronal (frontal) plane.** Bending of the neck in this plane is called **lateral flexion.** At certain joints, **rotation** also occurs about a longitudinal axis so that the face is turned to the right or left.

x

HEAD AND NECK

CHAPTER 1

GENERAL INTRODUCTION

TO obtain a useful idea of the important structures to be studied in the head and neck it is helpful to consider that in man they have the following functions:

(1) to contain and protect the brain. The exceptionally large size of the brain in man has produced a marked effect on the shape of the upper part of the head, the large frontal lobes giving it a rounded form.

(2) to contain the sense organs. For this purpose the head includes the nasal, the orbital and the auditory cavities. In order to direct the sense organs as required, it is essential that the head be mobile, allowing the orbits in particular to be turned in many directions.

(3) to chew and swallow food. The jaws with their teeth and muscles occupy a large part of the lower region of the head. The tongue, palate and pharynx provide a mechanism for swallowing food after it has been broken down and mixed with saliva. The food then passes into the oesophagus.

(4) to conduct the air towards the lungs. The air stream is moistened and warmed in the nasal cavities which are elongated chambers with folded walls, and then passed into the larynx and trachea.

(5) to carry the organs of speech. These include the larynx, the nasal and mouth cavities, palate, tongue, teeth and lips, which allow suitable adjustment of the air stream.

(6) expression. The facial muscles serve the important function of permitting facial expression which plays such a large part in life. Damage to the facial nerve controlling these muscles is socially a very serious accident.

DEVELOPMENT

The head still shows many signs that it is based on a segmental plan and that it once carried a number of branchial arches. The

1

segmented, myotomal musculature of the original front part of the head has become modified to form the eye muscles, and the oculomotor, trochlear, and abducent nerves supplying them represent ventral root nerves. The muscles which are derived from the myotomes behind the inner ear, and in the adult are found in the tongue, are innervated by the hypoglossal nerves (which are also ventral root nerves). The remaining musculature of the head is derived from the muscles of the branchial region, and the nerves that innervate it, trigeminal, facial, glossopharygeal, vagus and accessory, represent dorsal root nerves. In retaining motor as well as sensory functions these cranial dorsal roots are closer to the more primitive pattern than the spinal dorsal roots. In the latter the motor functions have been transferred to the ventral roots. Neck muscles, which move the head and vertebrae, are supplied by the spinal accessory and segmental spinal nerves.

The fish's head possesses gill arches separated by gill slits. In the human embryo the slits are never complete but evidences of the underlying pattern are seen in the pharyngeal (branchial) arches, with the pharyngeal clefts outside and the pharyngeal pouches inside. The primitive pattern is modified in man in the following manner. The 1st pharyngeal arch gives rise to the upper and lower jaws, the malleus and the incus (ossicles of the ear). The mandibular division of the trigeminal nerve supplies the 1st arch muscles which are almost entirely concerned with mastication. The auditory tube and middle ear represent the 1st pharyngeal pouch, separated by the tympanic membrane from the 1st pharyngeal cleft which persists as the external acoustic meatus.

The hyoid (2nd pharyngeal) arch contains cartilages that persist in man as the stapes (the 3rd ossicle), the styloid process of the skull and the upper part of the hyoid bone. The nerve of this arch is the facial and its muscles include the superficial muscles of the face used in expression. Part of the 2nd pharyngeal pouch contributes towards the formation of the tonsil.

The cartilages of the remaining arches contribute to the hyoid bone and the laryngeal cartilages. The glossopharyngeal nerve supplies the 3rd arch and the vagus the succeeding arches. Various portions of the 3rd and 4th pharyngeal pouches help to form important structures such as the parathyroid glands and the thymus.

2

The apparently irregular structure of the head can thus be better understood by considering how a regular pattern is modified to perform special functions.

ELEMENTARY OSTEOLOGY

The skull

The skull provides, (1) protection for the brain and sense organs, (2) attachment for the muscles that hold and move the head, (3) the skeleton of the jaws and the attachment for the muscles of mastication. It contains a series of cavities, namely the cranial cavity for the brain, the orbital, nasal and auditory cavities for the special sense organs, and the mouth. The floor of the brain cavity develops in cartilage and is called the **chondrocranium**. It becomes ossified to form the base of the skull and includes part of the cavities of the special sense organs. The vault of the skull and most of the bones of the face develop as membrane bone.

The proper development of the bones depends partly on hereditary factors and partly on the stresses to which they are subjected. For example, the round shape of the upper part is partly due to the outward pressure of the brain, and the full development of the mastoid process depends on the activity of the muscles attached to it and to the growth of its air cells. The shape and proportions of the face also change with the development of the teeth and air sinuses, and the mandible alters again in old age after the loss of the teeth. Many of the bones develop from separate centres of ossification and become united as growth proceeds. The adult skull is thus composed of a rather smaller number of bones than is found in the newborn baby or in other mammals such as the rabbit. The bones, with the exception of the mandible, are so firmly connected together in the adult that no movement takes place between them. At many of these articulations (sutures) the bony edges are separated by a thin layer of fibrous tissue.

It is much more profitable to study the whole skull than the separate bones. As a preliminary, however, it is necessary to determine the position and general shape of the individual elements before a detailed study of the skull can be appreciated. It is

3

essential that the student should refer to the articulated skull throughout the study of this section.

Figure 1 shows a sagittal section of the skull. First identify the body of the sphenoid bone in the midline. Then, passing round the skull, identify the ethmoid, frontal, parietal and occipital bones.

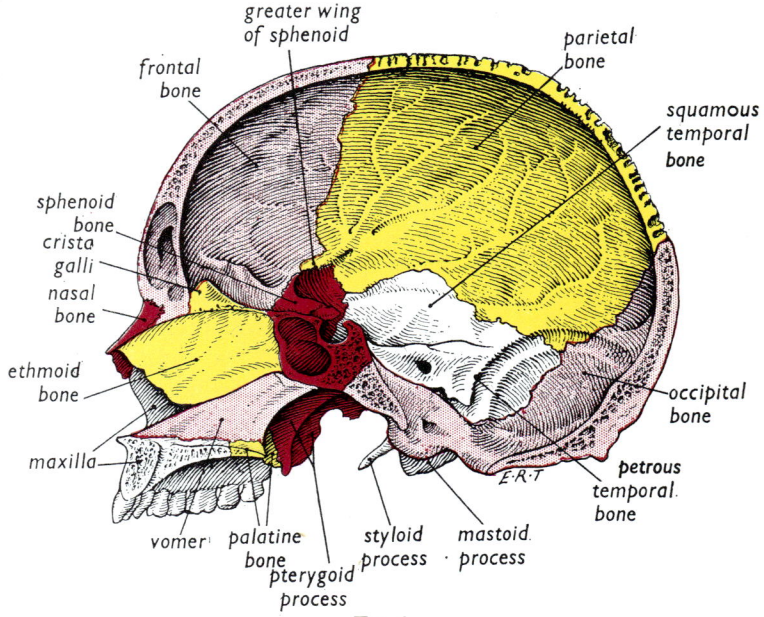

Fig. 1
The right half of a skull which has been sectioned sagittally.

The foramen magnum is in the occipital bone and anterior to the foramen is the basilar part of the occipital bone which fuses with the body of the sphenoid. The lateral wall of the vault of the skull is formed by the squamous portion of the temporal bone, the greater wing of the sphenoid, the frontal, the parietal and the occipital bones. The highest point of the vault is called the **vertex**.

Projecting downwards from the sphenoid bone is the pterygoid process, and from the temporal bone are the styloid and mastoid processes. In front of the pterygoid processes are the bones of the face. The ethmoid forms most of the roof and the upper part of the nasal cavity. The perpendicular plate of the ethmoid forms part

4

of the septum of the nose and projects up as the crista galli into the cranial cavity. The vomer helps to form the septum, and the nasal bones form the bridge of the nose. The maxilla is the largest bone of the face, and the alveolar process (with the teeth) and the palatine process (forming part of the hard palate between the nose and the mouth) can be seen. The back part of the hard palate is formed by the horizontal plate of the palatine bone. The mandible consists of a body (with the teeth) and two rami, each of which has a condylar process behind and a coronoid process in front (Fig. 10).

The structure of the skull bones

The bones formed from the chondrocranium are on the whole thick and hard, and the sphenoid and temporal contain cavities (air sinuses) communicating with the nose and pharynx respectively. The maxilla (of the face), the frontal (of the vault) and the ethmoid (of the nose) also contain air sinuses. The bones are covered with periosteum. Inside the skull the periosteum (**endocranium**), fuses with the dura mater covering the brain. In the air sinuses the periosteum fuses with their lining mucous membrane to form a mucoperiosteum. Over the vault of the skull, the periosteum forms the **pericranium**. The bones of the vault consist of inner and outer tables of compact bone with a cavity between containing red marrow and blood vessels (the diploë). The lining of the diploic cavity is the endosteum. Veins which connect venous channels inside and outside the skull are called **emissary veins** (Fig. 33).

The cervical vertebrae

The seven vertebrae in this part of the vertebral column are characterised by a foramen in the transverse processes. The 1st (the atlas) has no body and no spinous process but long transverse processes, and the 2nd (the axis) has the dens projecting up from the body behind the anterior arch of the atlas.

5

CHAPTER 2

THE BACK OF THE NECK AND THE TRUNK

INTRODUCTION

THE head is kept balanced on the vertebral column by the action of the neck muscles. Some of the muscles of the back of the neck play a part in transferring the weight of the upper limbs to the column. In animals where the head projects forwards it is supported by a powerful elastic **ligamentum nuchae** attached to the occipital bone and the cervical spinous processes. In man this ligament is small and gives attachment to muscles.

The vertebral column is a series of arches through which the weight of the head, trunk, and upper limbs is transmitted to the lower limbs. The thoracic and sacral (primary) curves are concave forwards and the cervical and lumbar (secondary) curves are convex forwards. The segmental nature of the column permits movements between adjacent vertebrae (flexion and extension about a transverse axis, rotation about a longitudinal axis and lateral flexion about an anteroposterior axis). Stability of the column as a whole depends on the intervertebral discs between the bodies, the numerous ligaments between the different parts of the vertebrae and the muscles of the back. There are also anterior muscles attached to the vertebrae in the cervical and lumbar regions and although there are no anterior muscles in the thoracic region, the anterior abdominal muscles are involved in movements of the vertebral column. The posterior vertebral muscles make an elaborate system; some of them run straight up and down, others obliquely in two main directions upwards and outwards (from lower spinous processes to higher transverse processes or ribs), and upwards and inwards (from lower transverse processes to higher spinous processes). This arrangement provides a set of braces able to hold up the weight of the head and trunk and yet allow movements. Normal activities frequently bring the line of weight (a vertical line through the centre of gravity of the segment of the body above the axis of the movement) in front of the vertebrae, and large back muscles are required to restore the normal upright position (Fig. 2). The muscles and joints of the

6

vertebral column come under special stress when the body is bent forwards and when lifting a weight by extension of the back. Usually the lower limbs are also bent and straightened but heavy weights should be lifted by straightening only the bent thighs and legs and not by straightening the flexed trunk.

FIG. 2

The large antigravity muscles in the trunk and leg are shown.

On the skeleton and on the cadaver identify the mastoid process of the temporal bone, the external occipital protuberance, the superior nuchal line, and the spinous processes of the 2nd and 7th cervical and all the thoracic vertebrae.

DISSECTION

If the upper limb has been dissected then the skin has been removed from the back of the trunk, neck and head. The muscles

7

of the shoulder girdle (trapezius, latissimus dorsi, etc.) have been removed and the erector spinae muscle exposed.

If the upper limb has not been dissected, incise the skin along the following lines: (1) from the external occipital protuberance laterally to the mastoid process; (2) from this point downwards and laterally towards the tip of the shoulder; (3) down the midline of the back as far as the level of the highest points of the iliac crests; (4) laterally from this point along the iliac crest. Turn the skin flaps laterally and expose the trapezius muscle medially and the sternocleidomastoid muscle laterally where they are attached to the occipital bone.

Clean the occipital attachment of the trapezius muscle (Fig. 3). While cutting it away from the bone, find the **greater occipital nerve** which runs upwards about 2 cm from the midline, with the **occipital vessels** on its lateral side (Fig. 4). Cut through the attachments of trapezius muscle to the spinous processes. Identify the latissimus dorsi muscle below and the rhomboid muscles above (Fig. 3) and detach these muscles from the spinous processes. Deep to the trapezius and sternocleidomastoid muscles identify the **splenius capitis** running upwards and laterally. Cut its upper attachment and turn it down, preserving the greater occipital nerve. Identify the narrow **longissimus capitis** muscle, running upwards and laterally, and medial to it the vertical fibres of the **semispinalis capitis.** Cut transversely through both these muscles near the occipital bone and turn them downwards.

(Although instructions for dissection follow, it is suggested that it may not be necessary to dissect the deep muscles of the back.)

The spinous process of the axis will now be prominent. Starting from it dissect out the **rectus capitis posterior major,** going upwards to the occipital bone, and the **obliquus capitis inferior,** going laterally to its attachment on the transverse process of the atlas (Fig. 5). Trace the greater occipital nerve round the inferior border of the inferior oblique. Divide transversely the belly of the rectus capitis posterior major and find deep and medial to it the **rectus capitis posterior minor** attached to the posterior tubercle of the atlas. Laterally, identify the **obliquus capitis superior** passing from the tip of the transverse process of the atlas upwards to the occipital bone. Expose the posterior arch of the atlas and above it

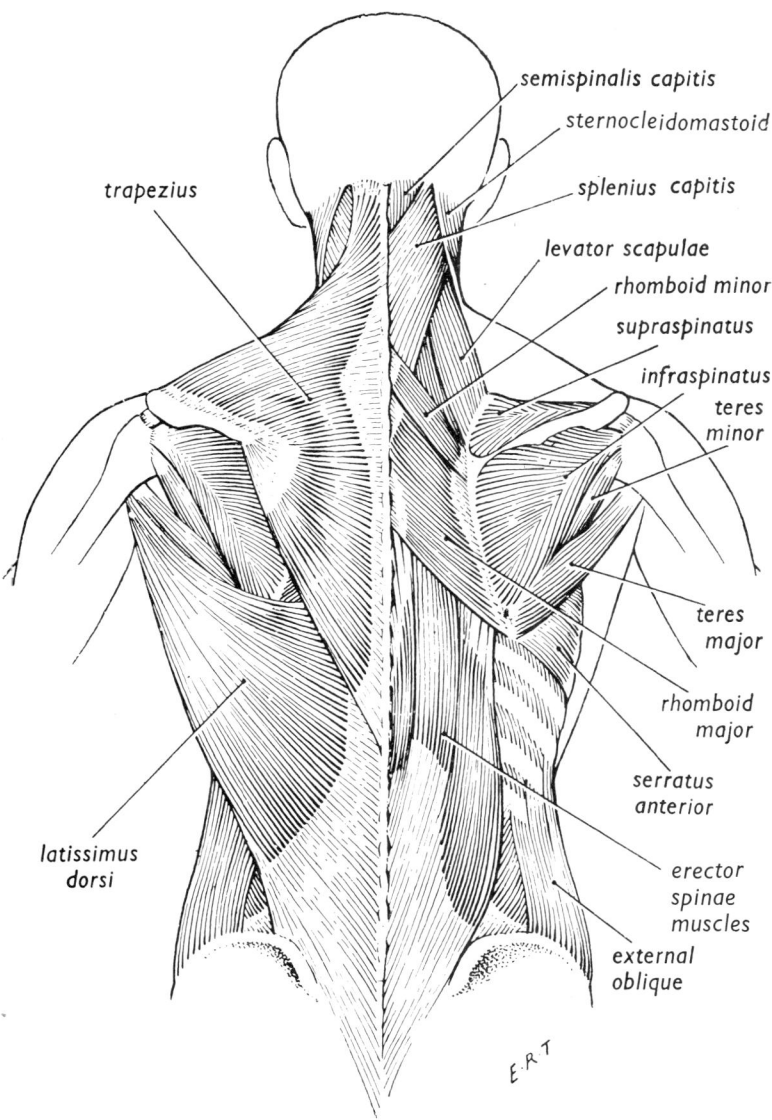

semispinalis capitis

sternocleidomastoid

trapezius

splenius capitis

levator scapulae

rhomboid minor

supraspinatus

infraspinatus

teres
minor

teres
major

rhomboid
major

serratus
anterior

latissimus
dorsi

erector
spinae
muscles

external
oblique

FIG. 3

The muscles of the back. On the right side, the trapezius and latissimus
dorsi have been removed.

9

identify the vertebral artery by incising the posterior atlanto-occipital membrane. The dorsal ramus of the 1st cervical nerve lies between the artery and the bone.

The **erector spinae** muscles of the vertebral column are arranged so that the long bundles are superficial and the short bundles more deeply placed. The muscles fill the hollow on each side of the midline between the spinous processes of the vertebrae and the angles of the ribs. These muscles are so interwoven with each other in some parts that individual identification is difficult and only the general arrangement need be known.

Superficially the muscles fall into three longitudinal groups; iliocostalis laterally, longissimus intermediately and spinalis medially (Fig. 6). All have attachments below to the fascia, ligaments and bone of the back of the sacrum and the posterior part of the ilium. The uppermost portion of the longissimus reaches the back of the lateral surface of the mastoid process (longissimus capitis). Cut across the superficial muscles and look for bundles of the more deeply placed muscles which are the semispinalis, multifidus, rotatores and intertransversarii (Figs. 6 and 7). The **semispinalis capitis** was seen passing to the occipital bone. The **levatores costarum** may be seen lateral to the erector spinae running downwards from the transverse processes to the corresponding ribs.

STRUCTURAL DETAILS

The occipital bone

The occipital bone develops partly in the base and partly in the vault of the skull, *i.e.* in cartilage and membrane respectively. About midway between the upper edge of the bone and the posterior margin of the foramen magnum there is a median elevation called the **external occipital protuberance**. This is easily palpable. The **superior nuchal line** extends laterally on each side from the protuberance towards the mastoid processes. It provides attachment for the trapezius medially and sternocleidomastoid muscle laterally. A ridge of variable prominence known as the **external occipital crest** extends from the external occipital protuberance to the foramen magnum; it provides attachment for the upper end of the ligamentum nuchae. The **inferior nuchal lines** extend laterally from the middle of the crest.

10

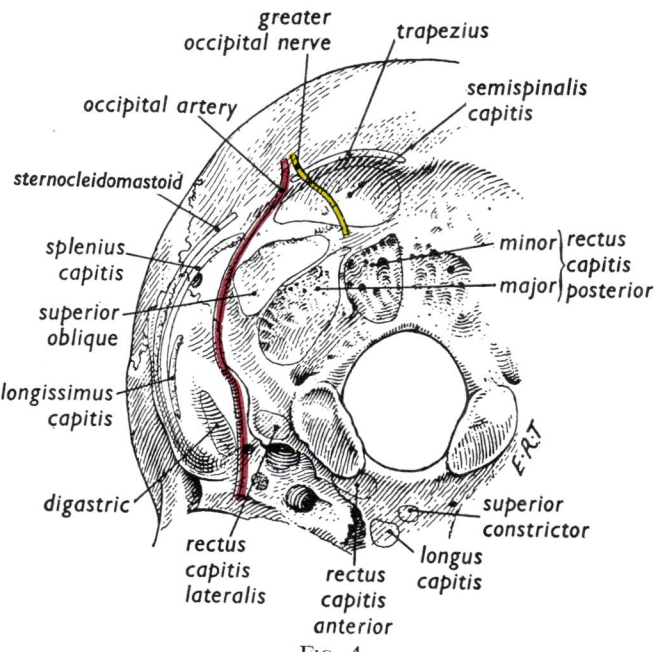

greater occipital nerve

trapezius

occipital artery

semispinalis capitis

sternocleidomastoid

splenius capitis

minor } rectus
major } capitis posterior

superior oblique

longissimus capitis

digastric

superior constrictor

rectus capitis lateralis

rectus capitis anterior

longus capitis

FIG. 4

The muscle attachments on the back of the base of the skull.

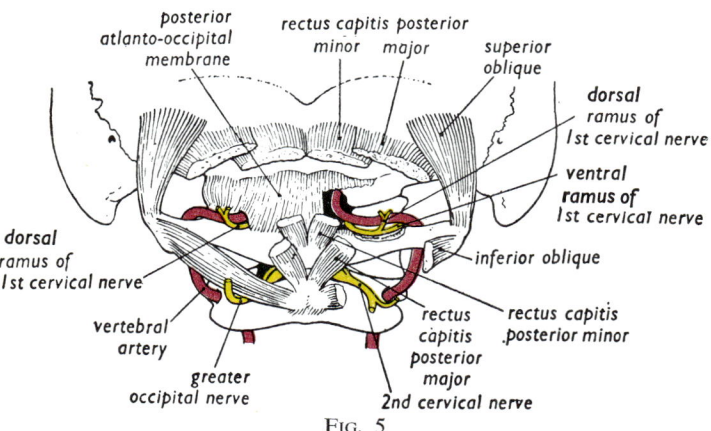

posterior atlanto-occipital membrane

rectus capitis posterior
minor   major

superior oblique

dorsal ramus of 1st cervical nerve

ventral ramus of 1st cervical nerve

dorsal ramus of 1st cervical nerve

inferior oblique

vertebral artery

rectus capitis posterior

rectus capitis posterior minor

greater occipital nerve

major

2nd cervical nerve

FIG. 5

The suboccipital region. On the right side the posterior atlanto-occipital membrane has been cut away to expose more of the vertebral artery and the 1st cervical nerve.

11

The cervical vertebrae

The cervical vertebrae have a foramen in each transverse process and the vertebral vessels run in this foramen except in that of the 7th. The spinous processes on the 2nd to the 6th are almost horizontal and their ends are bifid; there is none on the 1st and that of the 7th is not bifid. The bodies are small and the upper surface is " lipped " laterally and the lower surface is " lipped " anteriorly. The lateral lipping may form a synovial joint. The superior articular processes face backwards and upwards.

The **atlas** (1st) has an anterior and posterior arch but no body and no spine. Its upper articular facets for the occipital condyles are large, kidney-shaped and concave, the lower facets are flat and almost round. The transverse processes are long. The vertebral artery (Fig. 5) lies in a groove behind the **lateral mass**—that part of the atlas bearing the articular facets—and on the adjacent part of the upper surface of the posterior arch.

The **axis** (2nd) has the **dens** projecting upwards from its body. Developmentally this process is derived from the centrum of the atlas. The 7th vertebra resembles an upper thoracic vertebra in some respects. The spinous process is long and directed obliquely downwards.

Between the vertebral bodies are the **intervertebral discs** of fibrocartilage. The centre of a disc is semisolid (the **nucleus pulposus)** and the outer part is fibrous (the **anulus fibrosus**). Injury to or disease of these discs may later involve the spinal nerves. The spinous processes are joined together by strong **interspinous** and **supraspinous ligaments.** The **ligamenta flava** unite adjacent laminae and contain a large proportion of elastic tissue. The **ligamentum nuchae** is in the midline at the back and occupies the space between the external occipital crest above and the spinous processes of the 2nd-6th cervical vertebrae. In man it has many muscles attached to it but does not carry much of the weight of the head.

The joints of the neck will be considered later in Chapter 16.

The muscles of the back of the neck (Figs. 3, 4, 5, 6, 7 and 12)

The **cervical part of the trapezius** is attached above to the superior nuchal line and to the ligamentum nuchae, and below to the pos-

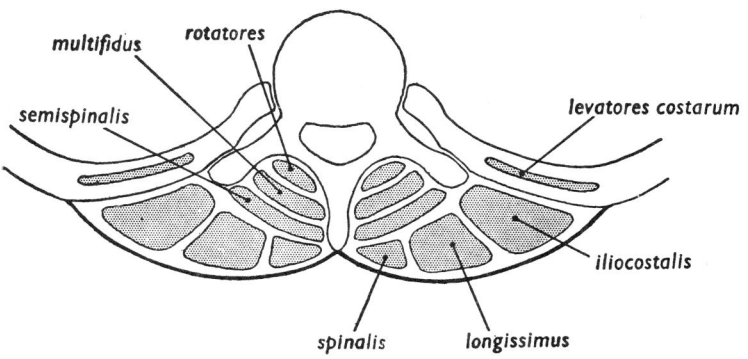

FIG. 6

Diagram of a transverse section of the erector spinae group of muscles.

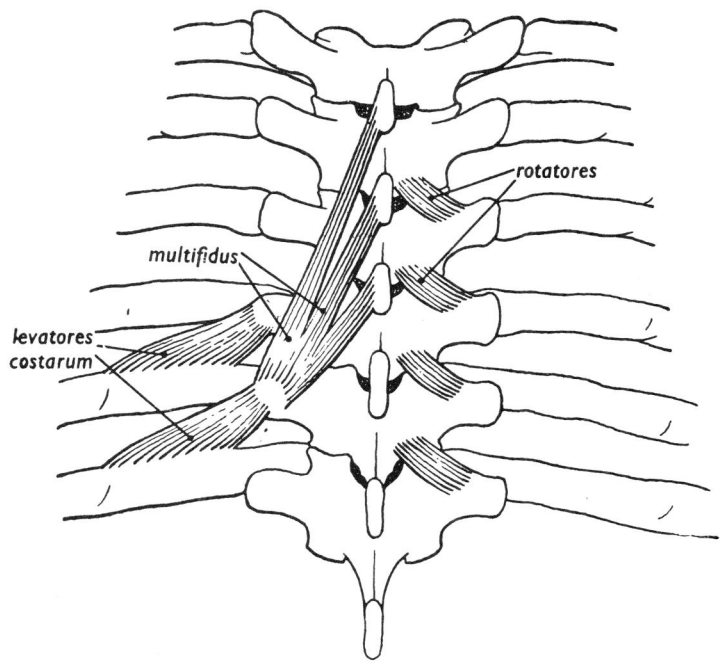

FIG. 7

The attachments of some of the smaller muscles to the transverse processes and the spinous processes of the thoracic vertebrae.

terior border of the clavicle laterally and the medial border of the acromion. It is supplied by the spinal accessory nerve and proprioceptor fibres from the ventral rami of the 3rd and 4th cervical nerves. With their clavicular ends fixed, both muscles extend the head. One muscle, acting with the sternocleidomastoid of the same side, bends the neck to the same side and turns the face towards the opposite side. The trapezius plays a large part in suspending the weight of the upper limb from the head and vertebral column.

The muscles of the back of the neck are supplied by the dorsal rami of the lower cervical and upper thoracic nerves. They extend the head and the cervical vertebrae, or turn the face to the same or opposite side, or assist in lateral flexion of the head and neck.

The small muscles of the suboccipital triangle are supplied by the dorsal ramus of the 1st cervical (suboccipital) nerve. They extend the head at the atlanto-occipital joints and rotate it at the atlanto-axial joints and are probably important in the positioning of the head for alignment of the two visual axes.

The muscles at the back of the neck prevent the head from falling forwards and in this way act against gravity. The muscles of the two sides similarly act to maintain the balance about the median plane. The erector spinae muscles are all supplied by the dorsal rami of the spinal nerves, branches of which will be found during the dissection, some going to the muscles and some to the skin.

The vessels and nerves of the back of the neck

The **occipital artery** arises from the external carotid artery, runs posteriorly and lies in the medial of the two grooves on the medial side of the mastoid process (Fig. 4). It then runs deep to the splenius capitis and superficial to the semispinalis capitis, pierces the aponeurotic attachment of the trapezius and ascends on the skull as high as the vertex, with the greater occipital nerve on its medial side.

The **vertebral artery,** a branch of the subclavian, ascends in the transverse foramina of the upper six cervical vertebrae (Fig. 5). The transverse foramen of the atlas is lateral to the corresponding foramen in the axis. The artery winds laterally towards the foramen transversarium of the atlas and then medially behind the lateral mass of the atlas. It then passes inferior to the edge of the posterior atlanto-occipital membrane which joins the posterior arch of the atlas to the posterior edge of the foramen magnum. The artery pierces the dura and arachnoid and enters the skull through the foramen magnum. The 1st cervical spinal nerve lies between the artery and the atlas and divides into a dorsal ramus which emerges below the artery, and a ventral ramus which passes forwards medial to it.

The **suboccipital plexus of veins** is found in this region. They communicate with the intracranial venous sinuses through the hypoglossal and condylar canals. The occipital veins communicate with the intracranial sinuses through the parietal and mastoid foramina.

The **cervical spinal nerves** give rise to dorsal rami which supply the back muscles and the skin of the back of the neck and scalp. The **greater occipital nerve** arises from the dorsal ramus of the 2nd cervical nerve. The ventral rami form the cervical and most of the brachial plexuses.

15

CHAPTER 3

THE SPINAL CORD AND ITS MEMBRANES

INTRODUCTION

THE spinal cord, with its accompanying nerves and coverings, lies in the vertebral canal. The walls of the canal consist of bone (bodies and neural arches), fibrocartilaginous discs (between the vertebral bodies), and ligaments (ligamenta flava between the laminae).

On the skeleton, identify the spinous processes, laminae, pedicles, and transverse and articular processes. Note the positions of the intervertebral foramina.

DISSECTION

With the handle of a scalpel remove the muscles from the spines and laminae of the vertebrae from the atlas to the 3rd lumbar and as far laterally as the articular processes. With a saw divide the laminae of the upper lumbar vertebrae close to the articular processes on both sides and cut across the ligamenta flava. Remove all the thoracic and cervical laminae and spinous processes with bone forceps. This will be simplified by flexing the trunk over a block. Examine the ligaments joining the spines (**supraspinous** and **interspinous**) and the laminae (**ligamenta flava**). The latter are composed of elastic tissue and extend as far laterally as the articular processes. Small veins pass through the ligamenta flava and connect the venous plexuses on the posterior surface of the laminae with the plexuses in the vertebral canal.

Within the vertebral canal a plexus of veins lies on the spinal **dura mater** in the **extradural space** which contains some fat. Note that each spinal nerve is surrounded by a sleeve of dura mater as it passes to the intervertebral foramen. Divide the dura vertically in the midline and turn the flaps laterally. The dura mater is firmly attached round the margin of the foramen magnum where it is continuous with the intracranial dura. Below this level the dura is separated from the walls of the vertebral canal by the extradural space. The **arachnoid mater** is a delicate membrane adherent to the inner surface of the dura and separated from the innermost layer, the **pia mater,** by the **subarachnoid space,** containing the **cerebro-spinal fluid.** The pia mater is a vascular layer closely investing the spinal cord and nerve roots.

The spinal cord begins at the level of the foramen magnum and ends at the level of the 2nd lumbar vertebra. A thin fibrous thread, the **filum terminale,** passes from the end of the cord to the back of the coccyx and, within the subarachnoid space, is surrounded by the lower lumbar, the sacral and coccygeal roots. These, together with the filum, are called the **cauda equina.** The subarachnoid space ends at the level of the 2nd sacral vertebra where the filum fuses with the dura. The spinal cord is the same length as the vertebral canal in the early embryo, but the bony canal rapidly outgrows the spinal cord. Identify the eight cervical, twelve thoracic, five lumbar and five sacral nerves. Note the increasing obliquity of their roots in the subarachnoid space.

Examine the spinal cord which is enlarged in two regions, (1) the **cervical enlargement,** associated with the nerve supply of the upper limb and (2) the **lumbar enlargement,** in the lower thoracic region of the vertebral column, for the supply of the lower limb. Below the lumbar enlargement the cord tapers to the **conus medullaris.** In the midline of the cord note the **posterior median sulcus** and lateral to it and extending the whole length of the cord the **posterior lateral sulcus.** Along this sulcus the fibres of the dorsal roots of the spinal nerves enter the cord. Follow one of the dorsal roots laterally and identify the **dorsal root ganglion** lying in the dural sleeve just medial to the junction of the ventral and dorsal nerve roots. The **posterior spinal arteries** can be seen on the surface of the cord near the posterior lateral sulcus.

Between the dorsal and ventral roots look for the **ligamentum denticulatum,** the scalloped edge of the pia mater which is attached to the dura. The lowest ligamentum denticulatum is usually crossed by the roots of the 1st lumbar nerve. In the upper cervical region note the spinal roots of the **accessory nerve** leaving the cord between the ventral and dorsal roots and joining together to form a trunk which passes upwards through the foramen magnum.

17

Above the arch of the atlas the vertebral artery pierces the dura mater and passes forwards and upwards between the pons and the basilar part of the occipital bone. A small portion of the cerebellum may be seen alongside the medulla oblongata in the foramen magnum.

Remove about 7 cm of the spinal cord with its membranes and nerves, and examine it in a dish of water. Incise the dura and arachnoid in front of the cord and find the **anterior median fissure.**

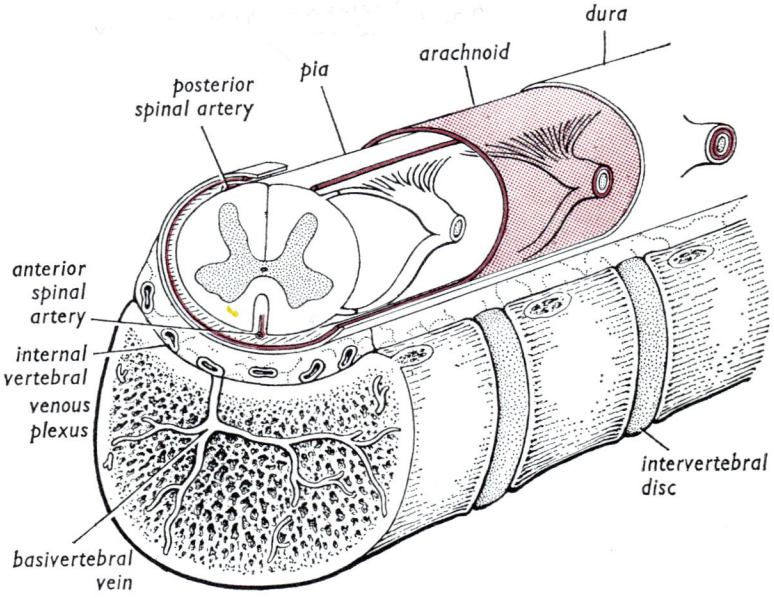

FIG. 8

Diagram showing some of the main relations of the spinal cord.

The **anterior spinal artery** (a fusion of branches from each vertebral artery) is found in the anterior median fissure. The fissure can be opened up easily because it contains an invagination of the pia mater. Anterolateral are the ventral nerve rootlets. Examine a freshly cut surface of the cord and note the white matter surrounding the **anterior** and **posterior horns** of the grey matter.

Remove the fat from the vertebral canal and expose the posterior longitudinal ligament covering the posterior aspects of the vertebral bodies and discs.

18

STRUCTURAL DETAILS

The meninges (Fig. 8)

The meninges form protective and vascular coverings for the spinal cord. The **extradural space** between the wall of the vertebral canal and the dura is filled with loose fatty areolar tissue and a considerable number of veins. These veins form longitudinal channels which anastomose with the spinal veins draining the cord, the basivertebral veins draining the vertebral bodies and the intervertebral veins passing through the intervertebral foramina. In the lower part of the extradural space the veins anastomose with the pelvic veins. The spinal **dura mater** is a strong, almost non-elastic tissue continuous at the edge of the foramen magnum with the inner layer of dura inside the skull. In shape, it is like a long tube, closed at its lower end. It has symmetrical lateral projections formed by the emerging spinal nerves. The dura of each lateral projection fuses with the periosteum of the intervertebral foramen and also becomes continuous with the outer fibrous covering of the nerves. The lower limit of the dura is about the level of the 2nd segment of the sacrum. The **arachnoid mater** is a thin membrane which lines the inside of the dura and is continuous with the **pia mater** on the surface of the cord. The arachnoid is separated from the dura by the **subdural space** containing a thin film of fluid. This space is continuous with lymph spaces in the sheaths of the spinal nerves. In a lumbar puncture a hollow needle is passed between the spinous processes of the 3rd and 4th lumbar vertebrae in the midline. The dura and the arachnoid are pierced and its tip then lies in the **subarachnoid space.** This space contains cerebrospinal fluid and the cauda equina. The fluid can be drawn off through the hollow needle. The cord ends at the 2nd lumbar vertebra and is therefore not endangered below this level.

The spinal arteries

The **anterior spinal artery** is formed in front of the medulla by the union of two branches derived from the vertebral arteries. It descends along the length of the cord in the anterior median fissure. The two **posterior spinal arteries** are branches of the posterior inferior cerebellar or the vertebral arteries. The anterior and posterior spinal arteries are supplemented by small branches passing through the intervertebral foramina and derived from the vertebral arteries and the segmental branches of the aorta.

The spinal cord and nerves

The spinal nerves come off the cord in a segmental pattern. There are usually eight cervical, twelve thoracic, five lumbar, five sacral and one coccygeal on each side. The 1st cervical roots are horizontal and the nerves emerge above the atlas. The subsequent roots pass downwards with an increasing obliquity.

CHAPTER 4

THE FACE AND SCALP

INTRODUCTION

THE muscles of the face are arranged round the orifices associated with the special senses—seeing, hearing, smelling and tasting. In man, their function in protecting or directing the orifices has, in the case of some muscles, been superseded by their use as a means of expression and communication. The thin skin with its rich blood supply aids these activities. Basically the muscles are arranged as sphincters with appropriate dilators. The scalp consists of specialised skin and underlying tissues covering the vault of the skull. The skin of the scalp and the face contains many hair follicles with their associated glands.

The front of the skull should be studied with the mandible in position, and with the **infra-orbital margins** in the same horizontal plane as the superior margins of the **external acoustic meati** (the Frankfurt plane). The skeleton of the face is irregular in contour and shows the orbital, nasal (piriform) and oral apertures. The maximum convexity of the frontal bone on each side is known as the **frontal eminence.** Above the **supra-orbital margins** there are well marked **superciliary arches** which approach the midline and form a medial elevation. The nasal bones are on each side of the midline inferior to this elevation.

The superior orbital margin is formed entirely by the frontal bone; at the junction of its medial third with its lateral two-thirds is the **supra-orbital notch** (or **foramen**) and through it the supra-orbital nerve and vessels emerge. The lateral orbital margin is formed by the frontal and zygomatic bones; the frontozygomatic suture, placed above its middle, can be felt as a depression in the living subject. The inferior orbital margin is formed by the zygomatic bone laterally and the maxilla medially. About 1 cm below its middle is the **infra-orbital foramen** from which emerge the infra-orbital nerve and vessels. The medial orbital margin is formed by the frontal process of the maxilla inferiorly, and by the frontal bone superiorly. The anterior nasal aperture is pear-shaped, and is formed by the nasal bones above and by the maxillae below and laterally.

The prominence of the cheek on each side is formed by the zygomatic bone. Below and medial to it is the maxilla. The mandible has a triangular elevation, the **mental protuberance,** in the midline near its lower border. The **mental foramen** is situated on each side below the interval between the two premolar teeth. Through this foramen emerge the mental vessels and nerve.

Viewing the skull from above, note :

1. the transverse **coronal suture,** between the frontal bone in front and the two parietal bones behind.

2. the longitudinal **sagittal suture,** between the two parietal bones. The coronal and sagittal sutures meet at the **bregma.** The skull vault is not completely ossified at birth and the region of the bregma has a diamond-shaped area occupied by fibrous tissue. This is known as the **anterior fontanelle** and usually closes by the eighteenth month.

3. the **lambdoidal suture,** separating the occipital bone behind from the parietal bones. The triangular area where these bones meet at the posterior end of the sagittal suture is called the **lambda.** In the newborn there is considerable separation of the bones and the space, occupied by fibrous tissue, is called the **posterior fontanelle.** This usually closes six to nine months after birth.

DISSECTION

Continue the midline incision through the skin from the external occipital protuberance over the top of the skull to the root of the nose, then down the middle of the nose to its tip. From this point, carry the incision round each side of the **nares** (anterior apertures of the nasal cavity) to the midline of the upper lip and down to the red margin of the lip (mucocutaneous junction). Incise right round the red margin to the midline of the lower lip and from there downwards across the point of the chin and down the midline of the neck to the sternum. Make elliptical incisions from the root of the nose just below the lower eyelid and above the eyebrow.

Remove the skin from the face and note that it is very thin in most parts and that the muscles of expression are attached

22

to it. Find the **parotid gland** below and in front of the ear and between the angle of the mandible and the mastoid process. It extends downwards over the sternocleidomastoid muscle for a variable distance. Its duct passes forwards from its anterior edge over the masseter muscle, bends medially and pierces the buccinator muscle. The branches of the **facial nerve** also pass forwards from the anterior edge of the gland (Fig. 11). They run to the forehead, cheek, upper lip, lower lip and neck. The middle branch runs almost horizontally between the lobe of the ear and the upper lip and is just below the duct of the parotid gland. The facial muscles are difficult to separate clearly because of the fat usually embedded in and between them. It is usually possible to define the orbicularis oculi, buccinator and some of the muscles of the lips (Fig. 9).

Find the **facial artery** as it crosses the lower edge of the mandible about 4 cm in front of the angle and trace it upwards as far as its terminal twigs at the inner angle of the eye. The **facial vein** runs posterior to the artery and is more superficial in its upper part and less tortuous.

Carefully remove the skin from the eyebrows and eyelids and find the **supra-orbital vessels** and **nerve** emerging above the eye about 2·5 cm from the midline. The **infra-orbital vessels** and **nerve** emerge from the maxilla about 1 cm below the infra-orbital margin and about 2·5 cm from the midline, and the **mental vessels** and **nerve** emerge from the body of the mandible about midway between its upper and lower borders and about 2·5 cm from the midline.

Over most of the face deep fascia is absent, but in front of the ear (over the masseter muscle and the parotid gland) it is present and is attached to the zygomatic arch. The temporalis muscle in the temporal fossa above the zygomatic arch is covered by thick fascia. Incise this fascia which is attached to the zygomatic arch and expose the muscle. Incise the fascia over the top of the skull and note that deep to it is a layer of loose areolar tissue. The handle of the scalpel can be passed forwards into the upper eyelid (Fig. 9), posteriorly as far as the superior nuchal line and laterally almost to the zygomatic arch. The **frontalis muscle** is embedded in the fascia over the forehead and the **occipitalis muscle** is found in the fascia above the superior nuchal line. Between the frontalis and occipitalis,

23

the fascia forms the **epicranial aponeurosis** and moves easily on the underlying bone. The **corrugator supercilii muscle** lies on either side of the midline above the eyebrows and deep to the frontalis; note that most of the fibres run laterally. Pull the eyelids laterally and define the **medial palpebral ligament** passing from the medial ends of the eyelids to the adjacent bone. Cut through the ligament and try to find the thin-walled **lacrimal sac** in the lacrimal fossa.

Remove the parotid gland piecemeal. Follow the branches of the facial nerve backwards and note that in the gland they lie superficial to the **retromandibular vein** which in turn is superficial to the **external carotid artery.** The retromandibular vein passes downwards and divides into anterior and posterior branches (Fig. 12). Follow the external carotid upwards behind the mandible. One of its terminal branches (the **superficial temporal artery)** passes upwards over the posterior end of the zygomatic arch into the scalp. The parotid gland extends medially and fills the space between the mandible anteriorly, the mastoid process posteriorly and the styloid process of the temporal bone medially.

STRUCTURAL DETAILS

The frontal bone

The frontal bone **ossifies** in membrane from symmetrical centres which fuse about the fifth year. It forms the skeleton of the forehead and the anterior part of the skull vault and extends backwards to the serrated coronal suture, where it articulates with the right and left parietal bones. It also possesses a pair of wide, thin horizontal shelves, forming the greater part of the roof of the orbits and the floor of the anterior cranial fossa. The **frontal air sinuses** are a pair of cavities between the tables of the frontal bone above the nose and the medial part of the orbits and they extend upwards, backwards and laterally for a variable distance. Their lining mucoperiosteum is continuous along the frontonasal duct with the mucous membrane of the nose.

The parietal bone

The two parietal bones, also membrane bones, form the skull vault posterior to the coronal suture. They articulate with each other at

24

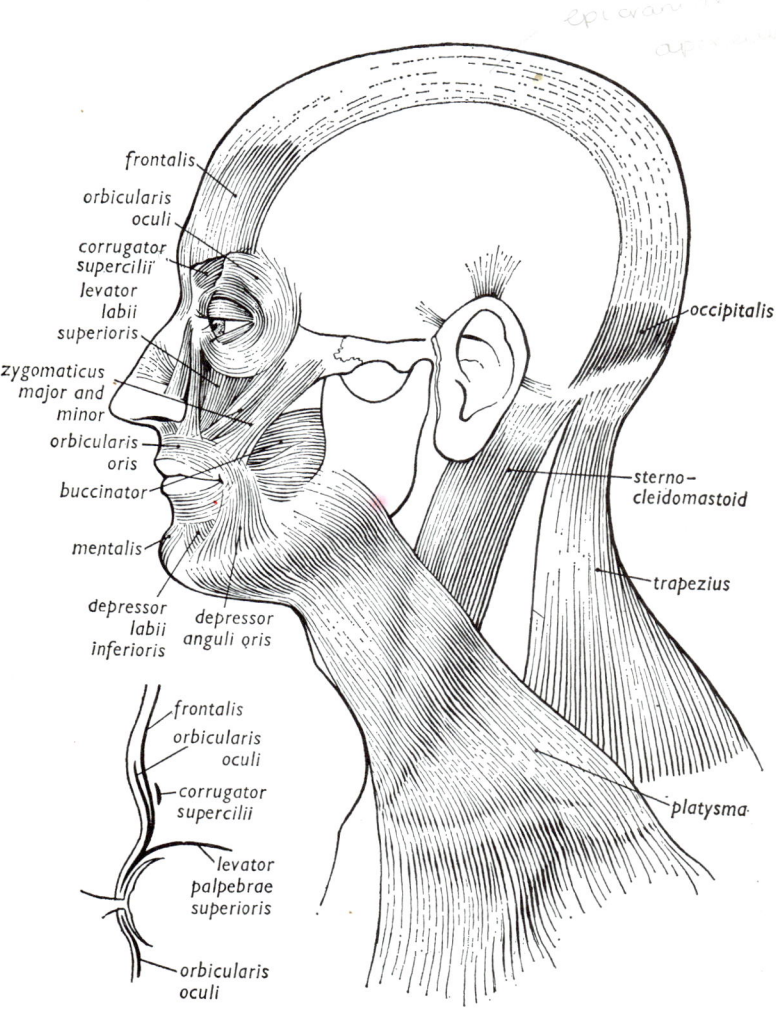

FIG. 9

The superficial muscles of the head and neck. The inset shows the relations
of the muscles in the upper eyelid.

the sagittal suture, posteriorly with the occipital bone at the lamb-doidal suture and inferiorly with the sphenoid in front and the temporal bone behind.

The maxilla

The two maxillae form the skeleton of the upper jaw, as well as participating in the formation of the hard palate, the floor and lateral wall of the nose, and the floor of the orbit. Most of the bone **ossifies** in membrane from a centre above the canine tooth. The part of the bone carrying the incisor teeth represents the **premaxilla** and has a separate centre of ossification. The suture between the premaxilla and maxilla may remain on the palatal surface until after puberty but its position is indicated by the **incisive fossa.** Each maxilla consists of a body and four processes. The body has a base lying medially and an apex directed laterally into its zygomatic process. Within the bone is a large cavity, the **maxillary air sinus** (the antrum). The base (nasal) surface of the disarticulated bone has an opening about 2 cm in diameter. In the articulated skull this opening is greatly reduced by the inferior concha, the lacrimal, ethmoid, and palatine bones. The posterior surface forms the anterior wall of the infratemporal fossa laterally and of the pterygopalatine fossa medially. The **pterygomaxillary fissure** lies between the posterior wall and the pterygoid process of the sphenoid. The superior (orbital) surface forms the greater part of the floor of the orbit. The anterior surface, forming the facial skeleton above the teeth, is concave and on it is the infra-orbital foramen.

The short, blunt **zygomatic process** extends laterally to articulate with the zygomatic bone. The **frontal process** extends upwards between the nasal and lacrimal bones to articulate with the frontal bone. The orbital surface of the frontal process has the vertical **anterior lacrimal crest** which is continuous with the inferior orbital margin. The part behind has a groove which, with a similar groove on the adjacent lacrimal bone, forms the hollow for the **lacrimal sac.** Inferiorly the **alveolar process,** the thickened portion carrying the teeth, forms with its fellow on the opposite side the **alveolar arch.** The roots of the teeth are embedded in its inferior surface which extends for a short distance beyond the last molar tooth as the **maxillary tuberosity.** The **palatine processes** of the two maxillae form the anterior two-thirds of the hard palate.

The zygomatic bone

The zygomatic (cheek) bone is a membrane bone and has a smooth lateral surface forming the hard part of the cheek. It forms the lateral part of the inferior margin of the orbit; posteriorly its articulation with the temporal bone completes the zygomatic arch; medially and anteriorly it articulates with the maxilla. A flattened process extends medially and separates the orbit from the temporal fossa. This is a characteristic feature of the primate skull.

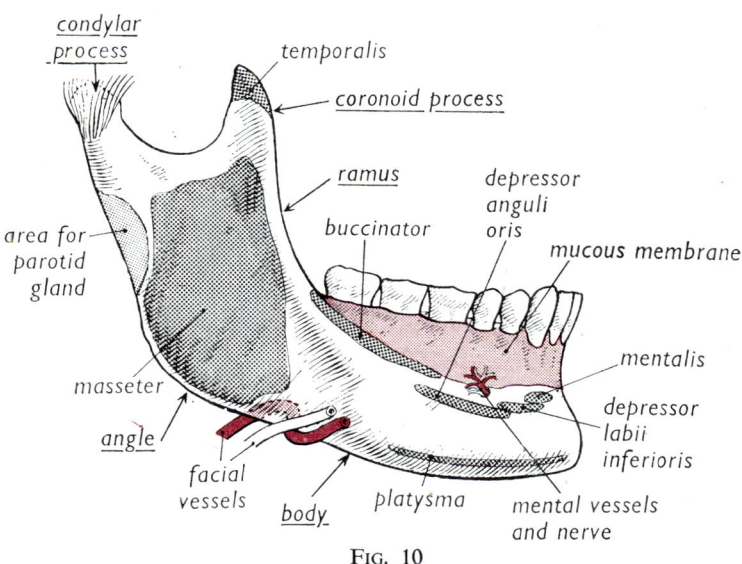

FIG. 10
The outer surface of the mandible.

The mandible (Figs. 10 and 19)

The mandible consists of an arched horizontal **body** with a vertical process, the **ramus,** projecting upwards on each side. The **angle** is where the lower border of the body meets the posterior border of the ramus. Two processes are seen projecting from the upper border of each ramus; the anterior one is the **coronoid process** for muscular attachment and the posterior one is the **condylar process,** which has an articular part called the **head.** This articulates

27

with the base of the skull. The narrow part of the condylar process inferior to the head is called the **neck.** The two processes are separated by the **mandibular notch.** The body has a smooth rounded inferior margin and an upper alveolar margin with sockets for the roots of the teeth. On the lateral surface is the **mental foramen** which, in the adult, is midway between the upper and lower borders, below the interval between the premolar teeth or about 2·5 cm from the midline. The vertical ridge in the midline expands below into a broad elevation, the **mental protuberance** that forms the prominence of the chin. (For development see page 53.)

The scalp (Fig. 33)

The skin and subcutaneous tissues over the vault of the skull are called the scalp. It is usually described as consisting of five layers.

1. The **skin** is thick and contains many sweat glands and hair follicles with their associated sebaceous glands.

2. The **connective tissue** is dense and contains many blood vessels and nerves. If the scalp is cut, the vessels are held open by their attachment to the connective tissue so that there is usually copious bleeding.

3. The **epicranial aponeurosis,** with the occipitalis and frontalis muscles, is firmly attached to the skin by the connective tissue layer. Behind, the occipitalis muscle is attached to the superior nuchal line; laterally the aponeurosis blends with the fascia over the temporalis muscle and is attached to the zygomatic arch. In front, the frontalis has no attachment to bone and its fibres merge with those of the orbicularis oculi.

4. Loose **areolar tissue** is found under the aponeurosis and allows the superficial layers to move easily on the deepest layer.

5. The **periosteum** (pericranium) covers the bone and is continuous with the endocranial (outer) layer of dura through the sutures and foramina.

The scalp has a rich arterial supply with many anastomoses between the larger branches. The veins drain into the extracranial jugular system but communicate with the veins of the diploic cavity of the skull bones and with the intracranial venous sinuses through the emissary veins.

The muscles of expression

These are found as a thin layer immediately under the skin of the face. They represent an extension of the muscles of the hyoid (2nd) arch and are innervated by the facial nerve. The mouth and eyes are surrounded by elliptical sphincter muscles and opposing radial dilator muscles, but around the nose and ear this arrangement is not so complete (Fig. 9).

The **buccinator** is an important muscle in the formation of the lips and cheek. It is attached to the outer surfaces of the maxilla and the mandible near the molar teeth, and to the pterygo-mandibular raphe between the hamulus (or hook) on the lower end of the medial pterygoid plate and the posterior end of the mylohyoid line on the medial surface of the mandible. The muscle passes forwards into the lips, the middle fibres decussating near the angle of the mouth. In the lips, the fibres blend with the orbicularis oris muscle and are reinforced by radially placed dilator muscles. Behind the pterygomandibular raphe, the buccinator is continuous with the superior constrictor muscle of the pharynx. The buccinator pushes the food between the teeth in chewing and is used in forced blowing. It will be further considered with the muscles of mastication.

The **orbicularis oris** surrounds the opening of the mouth. The fibres lying under the mucous membrane on the edge of the mouth purse the lips as in quiet whistling. The fibres attached to the maxilla below the nose and to the mandible produce marked pursing of the lips as in forced whistling. Above the mouth the **levator labii superioris** (used in sneering) passes upwards on the side of the nose to the maxilla and the **levator anguli oris** (used in smiling) also passes upwards to the maxilla. Below, the **depressor anguli oris** (expressing sorrow or dejection) runs downwards and outwards to fuse with the platysma of the neck and is attached to the edge of the mandible. The **depressor labii inferioris** and **mentalis** (used in pouting) are found nearer to the midline. Running backwards from the angle of the mouth superficial to buccinator is the **risorius** (used in grinning). It is attached to the fascia over the parotid gland. Passing upwards from the angle of the mouth to the zygomatic bone are the **zygomaticus major** and **minor muscles.** The **platysma** forms a sheet of muscle in the superficial fascia of the neck on each side

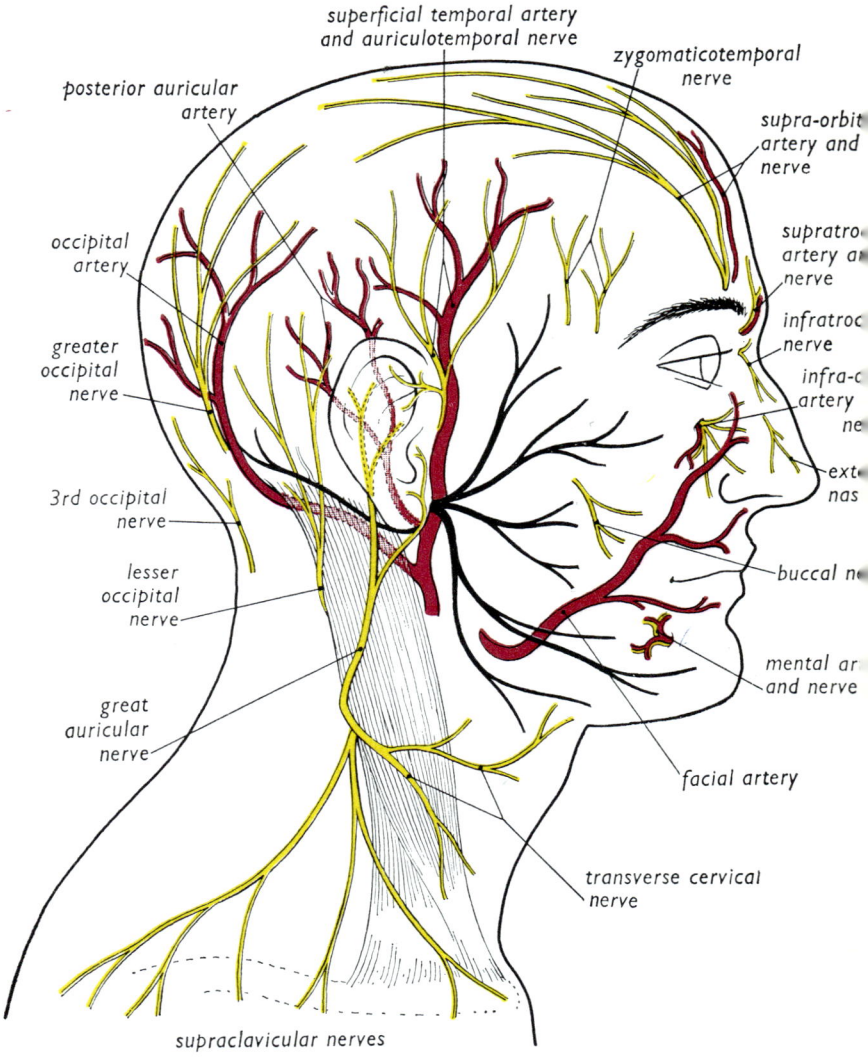

superficial temporal artery
and auriculotemporal nerve

zygomaticotemporal
nerve

posterior auricular
artery

supra-orbit
artery and
nerve

supratro
artery a
nerve

occipital
artery

infratroc
nerve

greater
occipital
nerve

infra-c
artery
ne

3rd occipital
nerve

ext
nas

lesser
occipital
nerve

buccal n

great
auricular
nerve

mental ar
and nerve

facial artery

transverse cervical
nerve

supraclavicular nerves

FIG. 11

The superficial nerves and arteries of the head and neck. The branches of
the facial nerve are shown in solid black.

30

of the midline. It continues downwards into the upper part of the chest, and above, it passes over the mandible and blends with the muscles surrounding the mouth.

The **orbicularis oculi** surrounds the opening of the orbit. The central fibres are in the lids and are responsible for winking and blinking, and the peripheral fibres pass over the superior and inferior margins of the orbit where they fuse with the frontalis above and the dilator muscles of the mouth below. The upper peripheral part of the muscle is used in frowning and when the eyebrows are pulled down over the eyes to protect them from the sun. When the whole muscle contracts, the skin is screwed up tightly and protects the eyeball. The orbicularis oculi is attached to bone medially and to the medial palpebral ligament. Elevation of the upper lid is produced by the **levator palpebrae superioris** which passes into the upper eyelid from the orbit. (It is supplied by the oculomotor nerve.) The **corrugator supercilii** is deep to the orbicularis above the medial end of the eyebrow.

The size of the nostrils can be controlled to some extent by the small muscles found attached to the cartilages forming the nasal aperture. The muscles attached to the auricle are vestigial.

The vessels of the face and scalp (Figs. 11 and 12)

The **facial artery** is a branch of the external carotid and crosses the lower border of the mandible about 4 cm in front of the angle. In the living subject a pulse can be felt here. The vessel has a very tortuous course upwards and forwards towards the inner angle of the eye. It runs in a plane between the superficial and deep facial muscles and in the face gives off branches to the cheeks, lips and muscles. It ends near the orbit by anastomosing with the ophthalmic artery which lies within the orbit. The small mental artery and the infra-orbital artery also supply the face. The scalp is supplied by the occipital, posterior auricular, superficial temporal, supra-orbital and supratrochlear arteries.

The **facial vein** begins at the inner angle of the eye, follows the course of the facial artery, and lies posterior and superficial to it. The vein is less tortuous than the artery. In front of the parotid gland, it joins the anterior branch of the retromandibular vein. The **retromandibular vein** is formed by the union of the superficial

31

temporal and maxillary veins in the parotid gland in the substance of which it soon divides into anterior and posterior branches. The anterior joins with the facial vein which is a tributary of the **internal jugular vein.** The posterior branch joins the posterior auricular vein and forms the **external jugular vein.**

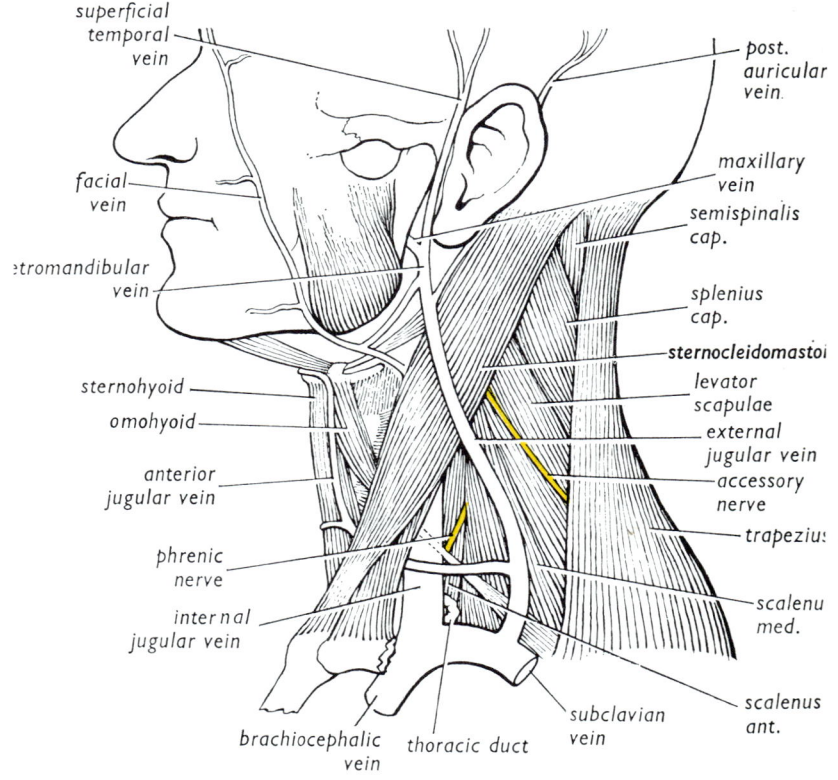

FIG. 12

The large veins of the face and neck are shown, also some of the muscles in the main triangles of the neck.

The nerves of the face and scalp (Figs. 11 and 59)

The **sensory** nerve supply of the face is derived mainly from the three divisions of the **trigeminal nerve—ophthalmic** above, **maxillary** in the middle and **mandibular** below. The

32

ophthalmic branches are the **supratrochlear, supra-orbital** and **lacrimal nerves** supplying the upper eyelid and the forehead from the medial to the lateral side. The upper part of the skin of the nose is supplied by the **infratrochlear nerve** and the lower part by the **external nasal,** which emerges at the lower border of the nasal bone and passes down the side of the nose. The maxillary branches are the **infra-orbital** (to the lower eyelid, the cheek, ala of nose and upper lip), the **zygomaticofacial** (to the skin over the zygomatic bone) and the **zygomaticotemporal** (to the skin over the zygomatic arch and the scalp above it). The mandibular branches are the **mental** (to the lower lip), the **buccal,** lying on the buccinator and supplying both the skin of the cheek and the mucous membrane of the mouth, and the **auriculotemporal** passing upwards in front of the ear behind the superficial temporal artery and supplying some of the skin of the auricle and scalp. The skin behind the auricle is supplied by the **great auricular** and **lesser occipital nerves** (ventral rami), and the **greater occipital** and **3rd occipital nerves** (dorsal rami.

The **muscles of expression,** including the platysma and buccinator, are supplied by the **facial nerve** which emerges from the stylomastoid foramen and runs forwards into the parotid gland where t divides into numerous branches to the face. Before entering the parotid gland, the facial nerve gives off branches to the stylohyoid, the posterior belly of the digastric and the occipitalis muscles. The branches to the face pass through the superficial part of the parotid gland.

The parotid gland

The parotid gland (one of the salivary group) occupies the very irregular space bounded behind by the external acoustic meatus, the mastoid process and the sternocleidomastoid muscle, in front by the mandible and its attached muscles, and medially by the styloid process and its attached muscles (Fig. 13). It extends upwards as far as the zygomatic arch, and forwards both deep and superficial to the ramus of the mandible. The gland is separated from the skin by the cervical fascia which above the angle of the mandible passes over the gland and fuses with the masseter muscle on the outer aspect of the ramus. The fascia is attached to the zygomatic

33

arch and as already described, becomes continuous with the temporal
fascia. There is no true capsule to the parotid gland and the lobules
of gland substance fill the available space and conform to the shape

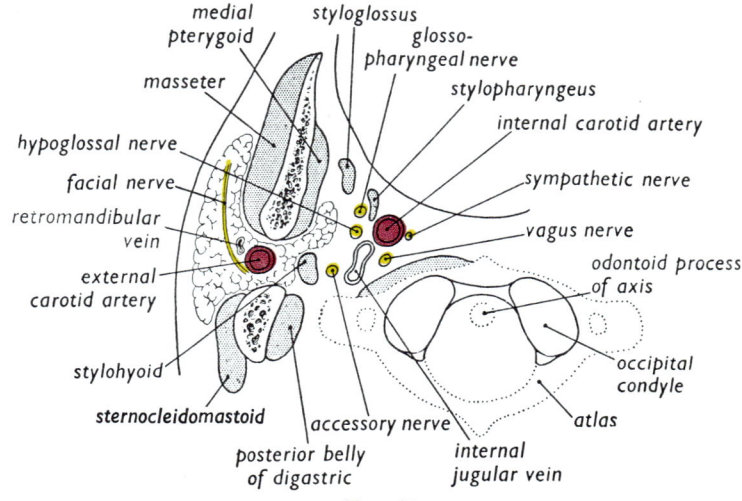

FIG. 13

Diagram of a transverse section through the parotid gland just below
the styloid process.

of the surrounding structures. The duct of the gland has already
been dissected and can be followed across the masseter to its anterior
edge, where it turns medially through the buccinator and opens on to
the mucous membrane of the cheek at the level of the upper 2nd
molar tooth.

Branches of the facial nerve pass through the substance of the
gland superficial to the retromandibular vein. The external carotid
artery enters its deep surface and divides about the level of the
neck of the mandible into the superficial temporal, which emerges
from the upper edge of the gland, and the maxillary which passes
medially out of the gland. Also in the substance of the gland are
numerous lymph nodes.

The anterior surface of the gland is in contact with the
mandible and the masseter and medial pterygoid muscles. The
posterior surface is in contact with the mastoid process, the
sternocleidomastoid and digastric muscles, and the external acoustic

meatus. Medially, the gland is separated by the digastric muscle from the internal jugular vein, and by the styloid process and stylohyoid muscle from the internal carotid artery. Above, the gland is in contact with the capsule of the temporomandibular joint. The postganglionic secretomotor nerve supply of the gland runs in the auriculotemporal nerve from the otic ganglion. The preganglionic fibres arise in the inferior salivary nucleus in the medulla oblongata, run in the glossopharyngeal nerve and its tympanic branch, and then in the lesser petrosal nerve to the otic ganglion (Fig. 32).

The lacrimal apparatus

Tears are formed by the lacrimal gland in the upper and outer side of the orbit and are secreted into the conjunctival sac. The tears, most of which evaporate, pass medially towards the **lacrimal puncta** which are the small openings of ducts on the summits of little papillae near the inner end of the free margin of the lids. From the puncta the upper **lacrimal canal** passes upwards and the lower downwards. Then they both turn medially and meet deep to the medial palpebral ligament. The single fused duct enters the upper part of the **lacrimal sac** which lies in the hollow formed by the lacrimal bone and the frontal process of the maxilla. The lower end of the sac leads into the **nasolacrimal duct** which passes to the inferior meatus of the nose. The nerve supply of the lacrimal gland is summarised in Fig. 57.

FUNCTIONAL ASPECTS

The face in the living subject

With the help of a mirror or on a partner, examine the effects produced by various facial and scalp muscles. It should be possible to feel the facial artery pulsating as it crosses the mandible and the superficial temporal artery as it crosses the zygomatic arch. The parotid duct is palpable as it crosses the contracted masseter. Pull the lower eyelid down and note how the conjunctiva is reflected from the lower eyelid on to the eyeball. It is similarly reflected from the upper eyelid on to the eyeball although this cannot be seen. Laterally the conjunctiva leaves the eyeball to pass on to the lateral parts of the eyelids. The conjunctiva thus forms a sac with an opening corresponding to the edges of the eyelids. At the medial angle there is a fold of conjunctiva, the **plica semilunaris,** best seen when the eye looks laterally, and a vascular (red) structure, the **lacrimal caruncle.** Near the medial end of the free border of each eyelid is the **lacrimal papilla** with its punctum. On the flattened free edge of the eyelid are the eyelashes, and the openings of numerous tarsal glands which are deep to the conjunctiva. Each eyelid contains a plate of dense fibrous tissue. Identify the mucocutaneous junctions of the eyelids and the lips.

CHAPTER 5

THE FRONT AND SIDES OF THE NECK

INTRODUCTION

ON the body, outline the clavicle and the lower border of the mandible from the midline in front to the angle behind. Identify the hyoid bone (posterior and slightly inferior to the mandible), the laryngeal prominence of the thyroid cartilage (Adam's apple), the cricoid cartilage (just below the thyroid cartilage) and the tracheal rings (Fig. 20). On the skull note the positions of the mastoid and the styloid processes of the temporal bone.

DISSECTION

Remove any remaining skin and superficial fascia from the neck. Deep to the platysma find the **external jugular vein** running from just behind the angle of the jaw to the middle of the clavicle. Find the **anterior jugular vein,** which drains the midline region anteriorly, passes laterally deep to the sternocleidomastoid muscle and joins the external jugular vein. Outline the position of the **sterno- cleidomastoid** and carefully remove the platysma and the superficial fascia from its surface. The sternocleidomastoid divides the neck into posterior and anterior triangles. The posterior is bounded behind by the anterior border of the trapezius, in front by the posterior border of the sternocleidomastoid and below by the clavicle. Note that its apex is posterior and its base anterior. The floor of the posterior triangle consists of muscles some of which were described on page 8. They are from below upwards the scalenus medius, levator scapulae, splenius capitis and semispinalis capitis. The first two will be seen more clearly later (Fig. 12 and page 65). The anterior triangle is bounded by the anterior border of the sternocleidomastoid, the midline of the neck and the lower border of the mandible. About halfway down the sternocleidomastoid muscle numerous cutaneous branches of the **cervical plexus** are seen emerging from its posterior border (Fig. 11). These are the **lesser occipital** (distributed to the region of the mastoid process), the **great**

37

auricular (to both sides of the ear and the parotid region of the face) and the **transverse cervical nerve** (to the side and front of the neck in the region of the thyroid cartilage). The **supraclavicular nerves** emerge somewhat lower down and supply the lower part of the neck and the upper part of the chest. The **spinal accessory nerve** runs downwards and backwards across the posterior triangle. The nerve emerges halfway down the posterior border of the sterno-cleidomastoid and disappears deep to the trapezius about 6 cm above the clavicle. The **greater occipital nerve** has been already dissected and traced to its area of supply on the back of the scalp. Clean all these nerves and note their distribution.

Note the attachments of the sternocleidomastoid and trapezius, above to the superior nuchal line and below to the clavicle. They are both enclosed by the deep cervical fascia. Cut across the sterno-cleidomastoid about its middle and turn its two halves upwards and downwards. The continuity between the posterior and anterior triangles can now be seen. In the lower part of the neck find the **omohyoid** muscle which has a central tendon, a superior belly attached to the hyoid bone and an inferior belly attached to the upper border of the scapula. Note the attachment of the central tendon to the medial end of the clavicle by a fascial sling; cut through the central tendon and trace the bellies upwards and downwards. Find the **internal jugular vein** lying deep to the superior belly of the omohyoid. Trace the vein upwards towards the angle of the jaw and downwards towards the sternoclavicular joint. It receives both the facial and lingual veins at about the level of the hyoid bone. Pull the vein laterally (or cut it if necessary) and expose the **carotid vessels** on its medial side. Starting below, trace the common carotid artery to its bifurcation at the level of the upper border of the thyroid cartilage and then identify the internal and external carotid arteries. The latter gives off a number of branches to the structures of the neck and head. Find the **vagus nerve,** lying between and behind the internal jugular vein and the carotid arteries. In the prevertebral fascia behind the carotid vessels find the **cervical sympathetic trunk,** which consists of a nerve with two or three ganglia on it. Medial to the common carotid artery identify below, the trachea and the oesophagus and higher up, the larynx and the pharynx. The oesophagus and pharynx are posterior

Remove the fascia from around the jaw and find the **sub-mandibular gland** deep to the body of the mandible, and the **posterior belly of the digastric muscle** inferior to the submandibular

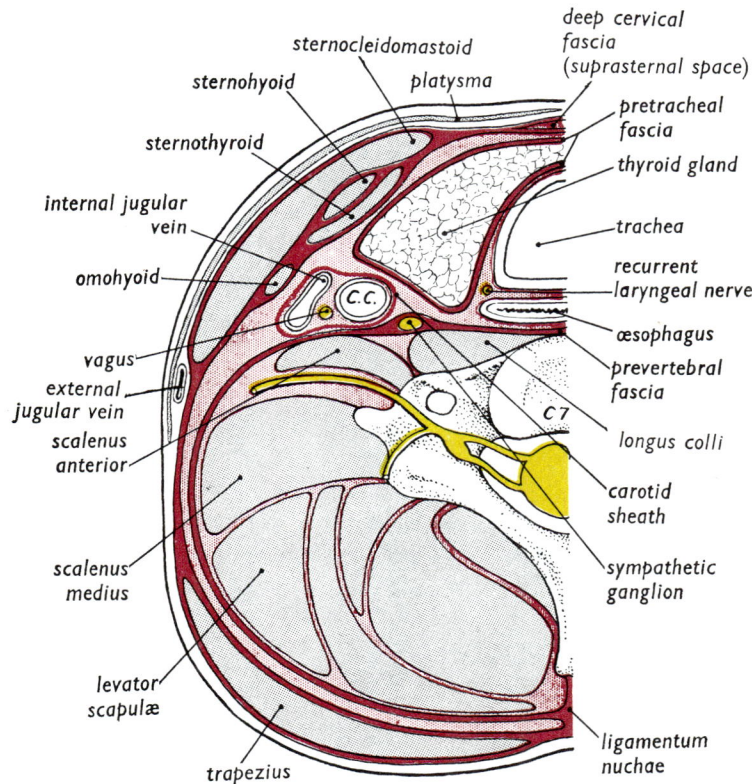

FIG. 14

Transverse section immediately above the 7th cervical vertebra. (C.C., common carotid artery.)

gland. Clean this muscle and the stylohyoid muscle which embraces the tendon of the digastric. The lower part of the gland should be pulled upwards and attached to the mandible. Cut through the facial vein and retract its ends. Find the hypoglossal nerve running downwards and forwards deep to the vein and superficial to the internal and external carotid arteries (Fig. 16).

Clean the external carotid artery and its branches. From below, the branches passing forwards are (1) the **superior thyroid** (passing downwards on the side of the larynx), (2) the **lingual** (coming off at the level of the hyoid bone and forming a loop upwards which is crossed superficially by the hypoglossal nerve), (3) the **facial** (disappearing deep to the submandibular gland). Passing backwards along the lower and upper borders of the posterior belly of the digastric are (4) the **occipital artery** and (5) the **posterior auricular artery** respectively. (6) The **ascending pharyngeal artery** arises just above the common carotid bifurcation and ascends on the lateral wall of the pharynx deep to the carotid sheath. The division of the external carotid artery into superficial temporal and maxillary arteries takes place behind the neck of the mandible in the parotid gland. Find the **ansa cervicalis** on the surface of the common and internal carotid arteries. This nerve loop is formed from a superior root, a branch of the hypoglossal nerve (actually a branch of the 1st cervical nerve) and an inferior root, a branch of the cervical plexus (from the 2nd and 3rd cervical nerves). The ansa and its roots supply infrahyoid muscles (Fig. 26). Lying medial to the internal carotid artery is the **superior laryngeal** branch of the vagus. It divides into an **internal laryngeal** branch which pierces the thyrohyoid membrane and an **external laryngeal** branch which runs with the superior thyroid artery.

STRUCTURAL DETAILS

The fascia of the neck (Fig. 14)

The **superficial fascia** of the neck consists of loose areolar tissue and in it are the platysma muscles and the superficial veins (the external and anterior jugular with their tributaries). The **cervical** (deep) **fascia** surrounds the neck and splits to contain the trapezius and sternocleidomastoid muscles on each side. The infrahyoid group of muscles is embedded in the deeper layers of this fascia. The deep fascia is attached to the lower border of the mandible and continues over the parotid gland to the zygomatic arch. The parotid and submandibular glands are ensheathed by this fascia and are separated from each other by a thickening called the **stylomandibular ligament** attached above to the tip of the styloid process and below to the angle of the mandible. From the deep fascia

40

a septum, the **prevertebral fascia**, passes medially in front of the muscles attached to the transverse processes of the cervical vertebrae and the nerves forming the cervical and brachial plexuses. The spinal accessory nerve in the posterior triangle lies between the cervical and prevertebral layers of fascia. The cervical sympathetic trunk is embedded in the latter and lies in front of the prevertebral muscles. In the midline of the neck are the pharynx and larynx (or trachea and oesophagus lower down) and their associated structures. These midline structures are covered by the **pretracheal fascia** which ensheaths the thyroid gland. Lying deep to the sternocleidomastoid, lateral to the midline structures and in front of the vertebrae, is a fascial compartment, which contains the main vessels and the vagus nerve. This is the **carotid sheath**, which is continuous below with the fibrous pericardium and is attached above to the base of the skull. There is a plane of cleavage between the carotid sheath and the prevertebral fascia so that when the sheath with its contents is drawn forwards, the sympathetic trunk remains with the prevertebral fascia.

The muscles of the neck

The **sternocleidomastoid muscle** is attached above to the mastoid process and the lateral part of the superior nuchal line. Below, it has a tendinous attachment to the manubrium and a muscular attachment to the medial third of the clavicle. As it passes upwards in the cervical fascia, it winds round the neck. At its lower end, it has medial and lateral borders, which become anterior and posterior above. When one sternocleidomastoid contracts it pulls the head to the same side and turns the face to the opposite side. When the muscles of the two sides work together they flex the head and the neck. Its motor nerve is the spinal accessory and the muscle also has a proprioceptive branch from the cervical plexus. The **trapezius muscle** has already been described on page 13.

The **digastric muscle** has two bellies and can raise the hyoid or depress the mandible. The **anterior** will be seen later to be attached to the deep aspect of the mandible at the side of the midline. The intermediate tendon is attached to the hyoid bone by a fibrous sling, and the **posterior** belly is attached to the lateral of the two grooves on the medial side of the mastoid process. The anterior belly is supplied by the mandibular division of the

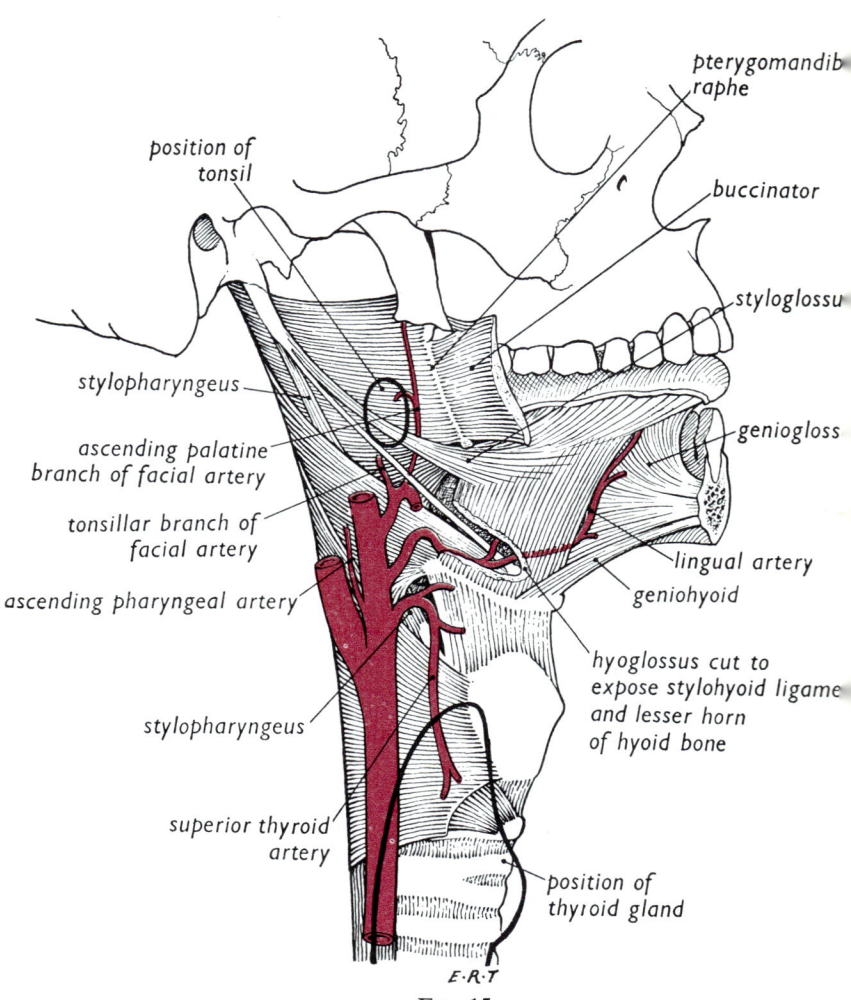

position of
tonsil

stylopharyngeus

ascending palatine
branch of facial artery

tonsillar branch of
facial artery

ascending pharyngeal artery

stylopharyngeus

superior thyroid
artery

pterygomandib
raphe

buccinator

styloglossu

geniogloss

lingual artery

geniohyoid

hyoglossus cut to
expose stylohyoid ligame
and lesser horn
of hyoid bone

position of
thyroid gland

E·R·T

FIG. 15

The relations of the carotid arteries to the pharynx are shown. The right lobe
of the thyroid gland and the tonsil are outlined in black.

trigeminal nerve through the mylohyoid branch of the inferior
alveolar nerve and the posterior belly by the facial nerve. The **stylohyoid muscle** raises the hyoid bone. It is attached above to **the**
styloid process and below to the hyoid bone, where it embraces
the tendon of the digastric. It is supplied by the facial nerve.

The carotid arteries

The **common carotid artery** arises on the right side from the brachiocephalic artery and on the left side from the arch of the aorta. It enters the neck behind the sternoclavicular joint from which it is separated by the brachiocephalic vein where the latter is formed by the internal jugular and subclavian veins. The common carotid divides into the external and internal carotids at the level of the upper border of the thyroid cartilage (Fig. 15). Medially are (1) the oesophagus and trachea with the recurrent laryngeal nerve between them below, and the pharynx and larynx above, (2) the thyroid and parathyroid glands and their vessels. The infrahyoid muscles overlap it from the medial side. Behind it are the vagus nerve in the carotid sheath and the sympathetic trunk in the prevertebral fascia on the prevertebral muscles. The omohyoid crosses the artery superficially. The internal jugular vein lies on the lateral side of the common carotid artery and becomes anterior to it behind the sternoclavicular joint. Covering both structures is the sterno-cleidomastoid muscle.

The **internal carotid artery** continues upwards in the line of the common carotid. It gives off no branches until it enters the carotid canal of the temporal bone. Coming off at the bifurcation of the common carotid is the **ascending pharyngeal artery,** which runs upwards on the lateral aspect of the pharynx, medial to the large vessels. The **carotid sinus** is a dilatation in the region of the bifurcation and lying medial to the sinus is the **carotid body.** The carotid sinus has in its walls nerve endings that are sensitive to changes in blood pressure and the carotid body is sensitive to chemical changes in the blood. The region is innervated by sensory branches of the glossopharyngeal nerve which produce reflex changes involving the heart and lungs through the stimulation of the cardiorespiratory centres in the medulla oblongata of the hindbrain.

The external carotid artery (Figs. 15 and 16)

The external carotid artery arises from the common carotid at the level of the upper border of the thyroid cartilage, runs upwards and divides into superficial temporal and maxillary arteries behind the neck of the mandible. The last part of the artery is embedded in the parotid gland. Its branches supply structures in the scalp, face

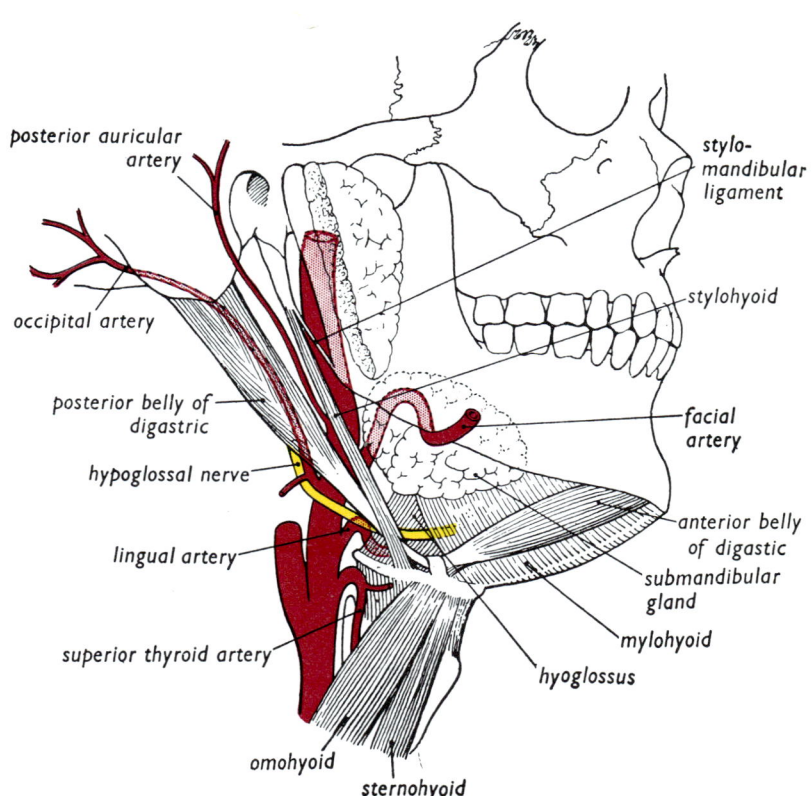

FIG. 16

Diagram indicating some of the relations of the parotid and submandibular glands.

and upper part of the neck. At first the external carotid lies antero-medial to the internal carotid but soon comes to lie lateral and superficial. Crossing superficial to the external carotid, from the angle of the jaw downwards, are the posterior belly of the digastric and the stylohyoid muscles, the hypoglossal nerve, and the lingual and facial veins. Along this part of its course it is overlapped by the sternocleidomastoid muscle. Passing between the external and internal carotids are the styloid process with its attached stylopharyngeus and styloglossus muscles, the glossopharyngeal nerve and the pharyngeal branch of the vagus (Fig. 17). Lying medially are the pharynx and the internal carotid artery. In the parotid gland, the retromandibular vein is superficial to the external carotid artery and the facial nerve branches are superficial to the vein.

Branches of the external carotid artery

1. The **superior thyroid artery** is the first large branch. It supplies structures below the hyoid bone down to the upper part of the trachea. Its superior laryngeal branch pierces the thyrohyoid membrane and supplies the larynx. Other branches are given to the thyroid gland and the infrahyoid muscles. The main artery is accompanied by the external laryngeal nerve.

2. The **lingual artery** arises opposite the hyoid bone, forms an upward loop on the middle constrictor muscle of the pharynx where it is crossed superficially by the hypoglossal nerve, disappears deep to the hyoglossus muscle and supplies the tongue.

3. The **facial artery** arises just above the lingual and after lying on the pharynx grooves the submandibular gland posteriorly where it may loop upwards as high as the superior constrictor and lie lateral to the tonsil. It gives off an important branch to the soft palate. It lies deep to the body of the mandible and winds round its lower edge in front of the attachment of the masseter where it is easily palpable. Its tortuous course on the face has been examined (page 31).

4. The **superficial temporal artery** was dissected with the face and the **maxillary artery** will be dissected later (Chapter 8).

5. The **occipital** and the **posterior auricular arteries** pass backwards below and above the posterior belly of the digastric respectively and supply the back of the scalp and upper part of the neck (page 14).

6. The **ascending pharyngeal artery** arises just above the bifurcation of the common carotid artery and ascends on the surface of the pharynx. It gives off an important tonsillar branch.

The hypoglossal nerve

The hypoglossal nerve supplies all the muscles of the tongue except the palatoglossus (Figs. 16, 17, 18, 20). It leaves the skull through the hypoglossal canal and lies medial to the internal jugular vein. It runs laterally behind the vagus and internal carotid artery and then between the internal jugular vein and the

45

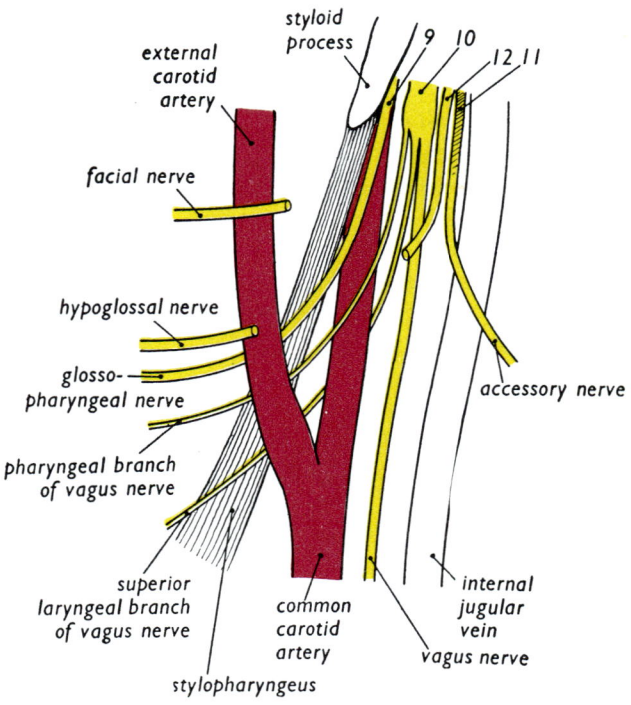

FIG. 17

Diagram of the relations of some of the cranial nerves to the internal and external carotid arteries.

internal carotid artery. It hooks round the occipital artery and crosses superficial to the internal and the external carotid arteries and the loop of the lingual artery on the middle constrictor of the pharynx. It lies deep to the posterior belly of the digastric and stylohyoid. It then becomes more superficial, passes forwards on the hyoglossus just above the hyoid bone, and enters the tongue deep to the mylohyoid. On the hyoglossus, the deep part of the submandibular gland, the submandibular duct and the lingual nerve lie above it. It gives off the superior root of the ansa cervicalis and branches to the geniohyoid and thyrohyoid muscles. These three branches are derived from the ventral ramus of the 1st cervical nerve (Fig. 26).

CHAPTER 6

THE MOUTH AND TONGUE

INTRODUCTION

BEFORE proceeding with the deep dissection of the neck it is necessary to remove part of the mandible. The opportunity will therefore be taken to examine the structures of the tongue and the floor of the mouth. The tongue is essential to the proper chewing of food because its keeps the food between the teeth. The muscles of the cheek (especially the buccinator) prevent the food accumulating in the **vestibule** of the mouth between the cheek and the jaws. The tongue also takes a very active part in the early stages of swallowing solid foods, the bolus of food being pressed against the hard palate first by the tip and then by the dorsum of the tongue. The lips, teeth and tongue modify the sounds produced by the larynx, thus making distinct, articulate speech possible. Because of its rich nerve supply the tongue is an important sensory organ, being especially associated with taste in the adult. In the newborn child, the lips and the tongue play an essential part in suckling.

The free part of the tongue consists of a mass of fine muscle fibres with very little fibrous tissue. These intrinsic muscles are connected by coarser extrinsic muscles to the mandible, the hyoid bone, the styloid process and the soft palate. Both groups of muscles change the shape and position of the tongue.

On the base of the skull (Fig. 18), identify the pterygoid process with its lateral and medial plates above and behind the hard palate, and the styloid process of the temporal bone. On the inner aspect of the mandible note the mental spines (the genial tubercles) lateral to the midline, and the mylohyoid line and groove passing backwards and upwards towards the mandibular foramen with its projecting lingula (Fig. 19). The hyoid bone consists of a body anteriorly, with two processes projecting backwards on either side, the lesser and greater horns, which may be cartilaginous in part. The lesser horn is above the greater and is connected to the styloid process by the stylohyoid ligament (Figs. 15 and 25).

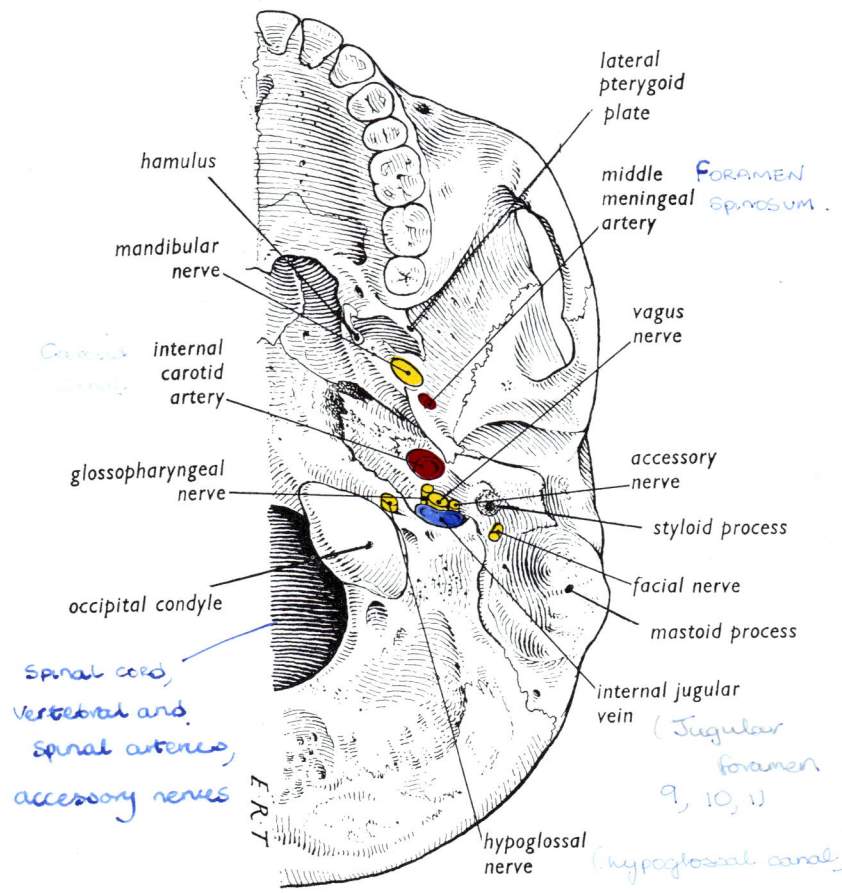

FIG. 18
The inferior surface of the base of the skull.

DISSECTION

Replace the submandibular gland and remove any cervical fascia between the mandible and hyoid bone. Running upwards and forwards is the anterior belly of the digastric muscle. Clean this, cut its attachment to the hyoid and turn the muscle upwards. Passing from the mandible to the hyoid bone is the broad sheet

48

of the **mylohyoid muscle,** which meets its fellow of the opposite side in a midline raphe between the mandible and the hyoid bone. The two muscles form a diaphragm across the floor of the mouth between the body of the mandible and the hyoid bone (Fig. 16). The mylohyoid has a free posterior edge and behind and deep to this is the **hyoglossus muscle.** The hyoglossus passes upwards from the greater horn of the hyoid to the side of the tongue, and the hypoglossal nerve lies on the muscle and then passes between it and the mylohyoid. The lingual artery can be seen passing deep to the hyoglossus to reach the tongue. Turning round the posterior border of the mylohyoid into the mouth are the deep portion of the submandibular gland and its duct. Cut the lip muscles from the front of the mandible and saw through the bone about 1 cm from the midline. Pull the lower border of the mandible upwards as far as is possible and clean the mylohyoid. Inferior to its attachment to the mandible is the mylohyoid nerve.

Cut through the cheek from the angle of the mouth towards the angle of the mandible. Turn the cheek down and examine the attachment of the depressor labii inferioris to the mandible by removing the mucous membrane from the vestibule and from between the lip and the gum (the labio-alveolar sulcus). Saw through the mandible just in front of the insertion of the masseter and turn the bone downwards exposing the sulcus between the gum and the tongue (the alveololingual sulcus). In the midline between the tongue and the floor of the mouth is the **frenulum** with, on each side, the **sublingual fold** lying over the **sublingual gland.** At the medial end of this fold, at the side of the frenulum, is the opening of the submandibular duct.

Note the extensive attachment of the mucous membrane to the mandible, and remove the membrane from the inner aspect of the mandible and from the side of the tongue. The fascial space between the mylohyoid and hyoglossus muscles will be opened. Trace the **lingual nerve** (above) and the **hypoglossal nerve** (below) forwards to the tongue (Fig. 20). Find the **lingual vein** running with the hypoglossal nerve and trace it to the internal jugular vein. Note that the lingual nerve crosses lateral to the submandibular duct and then passes upwards medial to it. On the hyoglossus, below the lingual nerve and attached to it, is the **submandibular ganglion.**

49

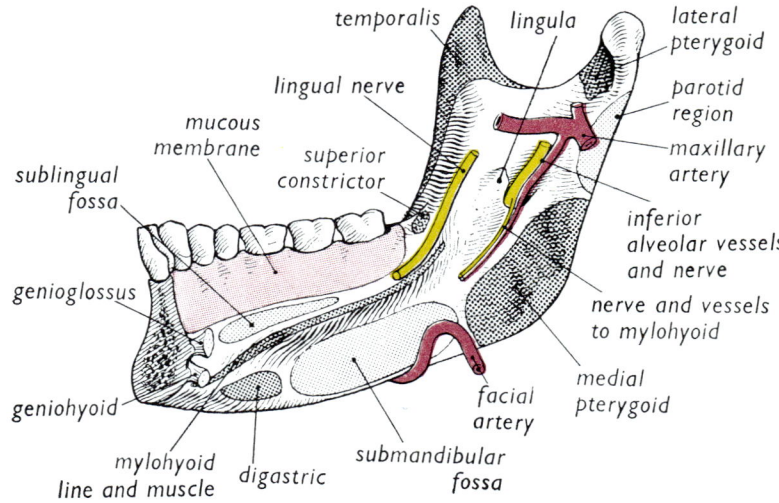

FIG. 19

The inner surface of the mandible. Inferior alveolar vessels and nerve
are shown entering the mandibular foramen.

Anterior to the hyoglossus, find the **geniohyoid** and **genioglossus
muscles** running from the mental spines to the body of the hyoid
and to the tongue respectively.

Trace the lingual artery forwards on the pharynx and incise
the hyoglossus parallel to the upper border of the hyoid bone thus
exposing the artery in the deeper part of its course towards the tip
of the tongue. Note the branch running along the upper border of
the hyoid bone, and other branches passing deeply towards the
upper surface (dorsum) of the tongue. At the posterior border
of the hyoglossus and deep to the styloglossus find the branches
of the **glossopharyngeal nerve** going to the base of the tongue. Trace
the nerve backwards and upwards and note how it passes lateral
to the stylopharyngeus muscle (Fig. 20).

Make a transverse incision across the " free " part of the tongue
(Fig. 21) and note the closely packed muscle fibres on either side of
a fibrous median raphe. Towards the back of the upper surface
of the tongue, find the pit of the **foramen caecum** lying in the midline behind the apex of the angle formed by the two rows of large

vallate papillae. This pit marks the lingual end of the embryological thyroglossal duct.

Examine the body of the mandible and note the mental foramen on its outer surface and, on the cut surface of the bone, the mandibular canal. Carefully detach the mylohyoid and free the bone. With a chisel, expose a length of the inferior alveolar nerve in the canal, and trace some of its branches towards the roots of the teeth (if present). Note the compact nature of the bone.

STRUCTURAL DETAILS

The mandible (Figs. 10 and 19, see also page 27)

The inner surface of the body presents behind the midline, two pairs of small elevations known as the **mental spines.** At the side of the mental spines near the lower margin is the **digastric fossa.** From below the mental spines, the **mylohyoid line** passes obliquely upwards and backwards as a prominent ridge. The sublingual salivary gland is in contact with the anterior part of the medial surface of the body of the mandible above the mylohyoid line. Its position is usually indicated by the smooth, shallow **sublingual fossa.** Below the mylohyoid line, and more posteriorly, there is another shallow depression, the **submandibular fossa,** in which the submandibular salivary gland lies.

The ramus is roughly rectangular with two surfaces, lateral and medial, and two borders, anterior and posterior. Its posterior border forms, with the inferior margin of the body, the **angle** of the mandible. The lateral surface is flat and provides attachment for the masseter muscle. The medial surface is irregular, and at its centre is the **mandibular foramen** leading to the **mandibular canal.** Anteromedial to the mandibular foramen is a spur of bone known as the **lingula.** The **mylohyoid groove** passes downwards and forwards from the lower margin of the mandibular foramen. The medial surface of the ramus below and behind the mylohyoid groove is roughened for the attachment of the medial pterygoid muscle. Another groove is often seen on the medial surface immediately postero-inferior to the 3rd molar tooth. The lingual nerve lies in this groove and may be injured during dental extractions.

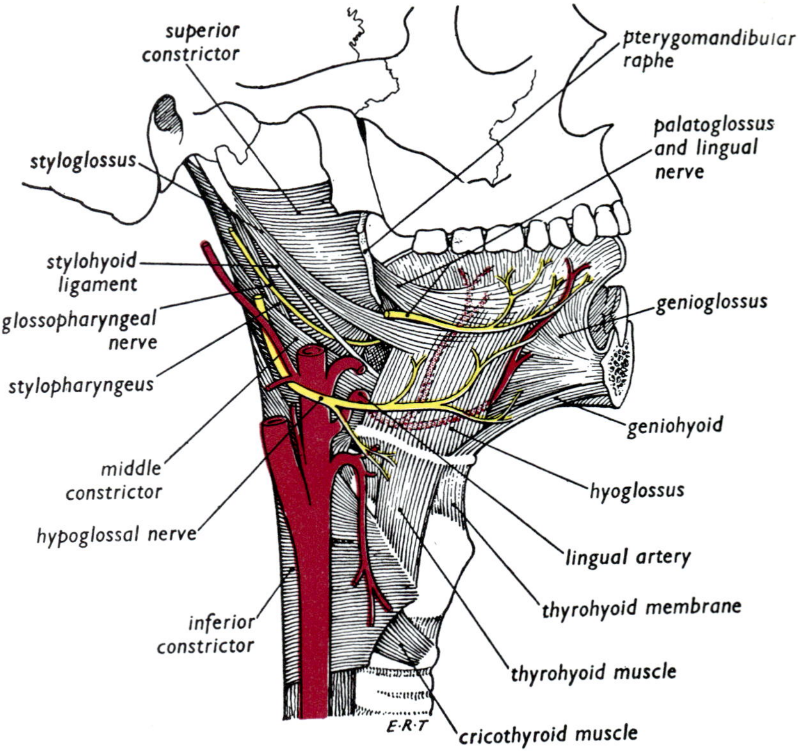

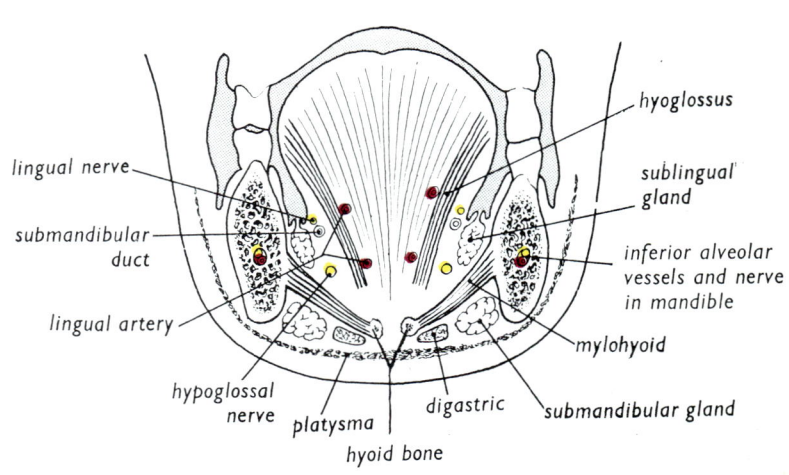

FIG. 20

The right half of the mandible has been removed to show the vessels and nerves entering the tongue.

FIG. 21

Diagram of a coronal section of the tongue indicating the muscle planes and the related structures. The hyoglossus muscle is in a posterior plane to the mylohyoid muscle.

The development of and postnatal changes in the mandible

Each half of the mandible starts to ossify at about the sixth week of intra-uterine life. It develops in the 1st pharyngeal (mandibular) arch by ossification in membrane on the outer surface of Meckel's

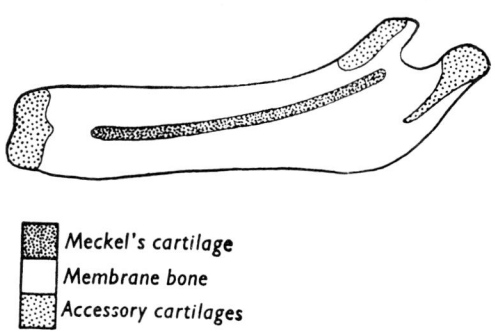

▨ Meckel's cartilage
☐ Membrane bone
▦ Accessory cartilages

FIG. 22

Development of the mandible. Membrane bone spreads from the surface of Meckel's cartilage and later invades the accessory cartilages.

cartilage. Accessory cartilages are found later in the regions of the midline, the coronoid process and the condylar process. They are ossified from the adjacent membrane bone. At birth the two halves are joined together anteriorly by fibrocartilage (symphysis menti). The inferior border of the body is poorly developed, and the mental foramen is situated nearer the lower than the upper border. The angle is much more obtuse than in the adult. The eruption of teeth results in deepening and thickening of the body and early in the second year the region of the symphysis menti ossifies. Bone is laid down on the inferior border of the body and the angle becomes less obtuse so that by the fourth year it is about 140°. During the whole period of growth elongation of the body is produced by absorption of bone at the anterior border of the ramus and by deposition at its posterior border. This lengthening allows room for the additional and larger permanent teeth. The ramus increases in height by growth in the region of the coronoid and condylar cartilages. The latter cartilage is present until about 25 years.

In the adult, the mental foramen is midway between the upper and lower borders of the body and the mandibular angle measures about 110°. As the teeth are shed with advancing age the alveolar process becomes absorbed so that the mental foramen lies nearer the upper border. At the same time, the angle becomes gradually more obtuse again till it is about 140°.

The teeth

Man and most mammals possess two sets of teeth, which erupt at different times during early life. The **deciduous (milk) teeth** of man appear between the sixth and twenty-fourth months and are gradually replaced by the **permanent teeth,** which appear between the sixth and twenty-fourth years. According to their position and shape, the teeth are named from the front, incisors (I), canines (C), premolars (P) and molars (M).

The dental formula for each half of each jaw is as follows:

Deciduous	I	C	M	
	2	1	2	

Permanent	I	C	P	M
	2	1	2	3

The permanent premolars replace the milk molars and the three permanent molars lie behind the premolar teeth.

Each tooth consists of a **root** or **roots** embedded in the jaw, a slightly constricted **neck** and a protruding part called the **crown.** The variations in the shape of the crowns and in the number of roots enable us to recognise individual teeth.

Incisor teeth have sharp cutting edges, used for biting food, and a single root.

Canine teeth have a large conical crown and also a single root

Premolar teeth, used for chewing, have two tubercles on the crown, hence " bicuspids ", and the tubercle on the labial side is more prominent. They have one root, except upper 1st premolars which have two.

Molar teeth, also used for chewing, have a large crown with three to five tubercles. The lower molars have two roots and the upper molars have three.

The normal times of eruption of teeth

A considerable delay in tooth eruption is commonly found in undernourished children and in those suffering from certain diseases. It is therefore advisable to know the average eruption times, although it must be appreciated that individual variation occurs. The approximate times of eruption are as follows :

	I	C	M	
Deciduous	7 8	18	12 24	months
	6 9	18	12 24	

	I	C	P	M	
Permanent	7 8	12	9 10	6 12 18-25	years
	7 8	12	9 10	6 12 18-25	

The permanent teeth of the lower jaw erupt slightly earlier than those of the upper.

Note especially 1. the first deciduous tooth to appear and its time of eruption, 2. the first permanent tooth to appear and its time of eruption, 3. the time of eruption of the third molar (wisdom) teeth, 4. one or more third molars may not erupt.

The hyoid bone (Figs. 15, 16 and 25)

The hyoid is a U-shaped bone lying almost horizontally in the upper part of the neck. It is developed from parts of the cartilaginous bars of the 2nd (hyoid) and 3rd pharyngeal arches. It consists of a central **body** and two **greater** and two **lesser horns.** The body is palpable in the neck at the level of the lower border of the mandible and from its lateral extremities the greater horns extend backwards and slightly upwards at the side of the neck. The lesser horns are small upward projections from the junction of the body with the greater horns.

The hyoid is connected inferiorly to the thyroid cartilage by ligament and muscle, and to the sternum and scapula by the **infrahyoid** muscles. Superiorly it is attached to the styloid process by a ligament and a muscle, and to the tongue, mandible and base of the skull by the **suprahyoid** muscles.

The muscles of the floor of the mouth and tongue

The **digastric muscle** (anterior belly) is attached to a small fossa lateral to the mental spines on the inner aspect of the mandible. The **mylohyoid** arises from the mylohyoid line on the inner aspect of the body of the mandible and the fibres pass downwards and medially to be attached to the hyoid bone and to the mylohyoid of the opposite side by a midline raphe. Both the mylohyoid muscle and the anterior belly of the digastric are supplied by the mylohyoid branch of the inferior alveolar nerve. The two muscles form the major portion of the floor of the mouth and when they contract the hyoid is raised, so carrying the tongue upwards towards the roof of the mouth. This happens during swallowing. The **geniohyoid** passes from the inferior mental spine to the body of the hyoid bone. If these muscles contract with the hyoid fixed by the infrahyoid muscles, the jaw is depressed.

The hyoglossus, genioglossus and styloglossus are all extrinsic muscles of the tongue. (The palatoglossus will be dissected later.) They form the main mass of muscle in the posterior part of the tongue (Figs. 20 and 21). The intrinsic muscles are difficult to dissect but they are made up of vertical, transverse and longitudinal muscle bundles. The **styloglossus** arises from the styloid process and passes forwards into the base of the tongue. Its fibres merge with the upper part of the hyoglossus. The **hyoglossus** arises from the greater horn and body of the hyoid bone, passes upwards and forms the main mass of muscle on the side of the tongue. It has important relations that have already been seen (Fig. 20). The **genioglossus**, attached to the superior mental spine, lies on the side of the frenulum of the tongue, and goes to the whole of the tongue as far forwards as the tip and backwards to the hyoid bone.

The nerve supply of the tongue and the floor of the mouth

The **hypoglossal nerve,** mainly a motor nerve but probably containing some proprioceptive fibres, supplies all the muscles of the tongue except the palatoglossus which is supplied by the pharyngeal plexus (page 136). Running in the hypoglossal nerve are fibres from the 1st cervical ventral ramus. These supply some of the infrahyoid muscles through the superior root of the ansa cervicalis,

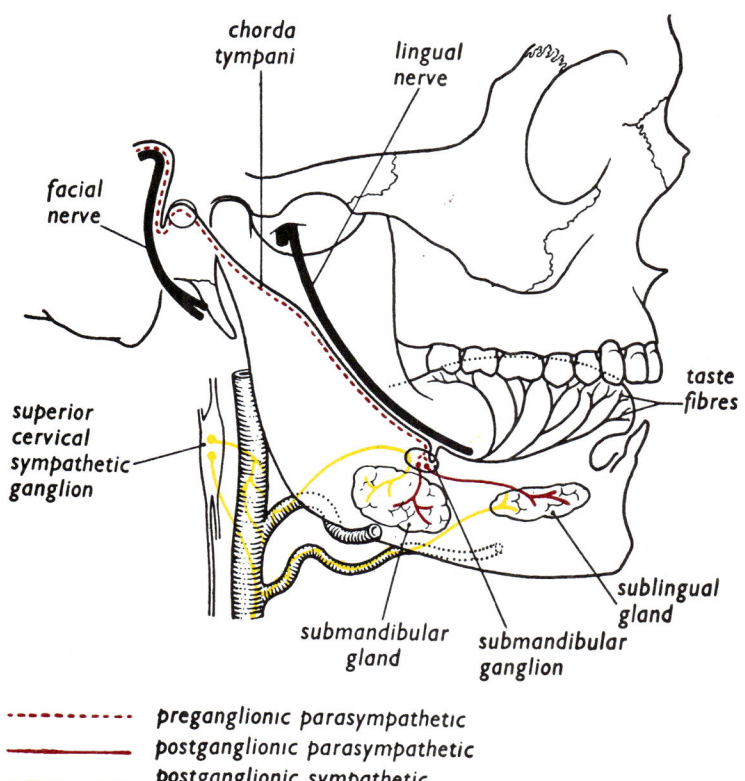

chorda tympani

lingual nerve

facial nerve

taste fibres

superior cervical sympathetic ganglion

sublingual gland

submandibular gland

submandibular ganglion

---------- preganglionic parasympathetic
——————— postganglionic parasympathetic
——————— postganglionic sympathetic

FIG. 23

Diagram of the position and connexions of the submandibular ganglion. The preganglionic secretomotor fibres arise in the superior salivary nucleus near the facial nucleus. The postganglionic fibres supply the submandibular and the sublingual glands.

and also the thyrohyoid and the geniohyoid muscles. The hypoglossal nerve emerges from the skull through the hypoglossal canal (anterior condylar foramen). It lies deep to the carotid sheath and then passes downwards between the internal jugular vein and internal carotid artery. About the level of the angle of the mandible it emerges between the vein and artery, and passes forwards over the internal and external carotid arteries and the loop of the lingual artery. It then lies superficial to the hyoglossus muscle where it is related to the deep part of the submandibular gland, the lingual nerve and the submandibular ganglion.

The **lingual nerve** is a branch of the mandibular division of the trigeminal nerve. It is the sensory nerve of the anterior two-thirds of the tongue and of the mucous membrane of the mouth between the tongue and the mandible including the inner side of the gum. A branch of the facial nerve, the **chorda tympani,** joins the lingual nerve near the base of the skull and contains (1) secretomotor and vasodilator fibres to the submandibular and sublingual glands, and (2) taste fibres from the anterior two-thirds of the tongue (Fig. 23). The fibres to the salivary glands are parasympathetic, travel in the facial, chorda tympani and lingual nerves, and synapse in the submandibular ganglion. The preganglionic parasympathetic fibres have their cell bodies in the superior salivary nucleus in the medulla oblongata. The chorda tympani leaves the facial nerve during its course in the temporal bone and comes out of the skull through the petrotympanic fissure posterior to the mandibular fossa of the temporal bone. The chorda tympani joins the lingual nerve deep to the lateral pterygoid muscle.

The **glossopharyngeal nerve** emerges through the jugular foramen and descends along the posterior border of the stylopharyngeus muscle which it supplies. Together with the muscle and the **pharyngeal** branch of the vagus, it passes between the internal and external carotid arteries, crosses superficial to the stylopharyngeus and then deep to the stylohyoid ligament to reach the back of the tongue. It is a sensory nerve at this level and carries general sensation and taste from the posterior third of the tongue. It gives a sensory branch to the pharyngeal plexus. It also contains preganglionic parasympathetic fibres which eventually reach the otic ganglion (page 35 and Fig. 32).

The mucous membrane of the tongue

The mucous membrane is thin on the floor of the mouth and the sides of the tongue, but is thicker and rougher over the dorsum of the tongue where numerous papillae can be seen. Here it also contains taste buds and there are collections of lymphoid tissue, especially at the back. The large **vallate papillae** are easily recognised and form a "∧" whose limbs pass back to the apex immediately in front of the **foramen caecum.** The taste buds of the vallate papillae and the mucous membrane behind are supplied by the glossopharyngeal nerve. Those in front are supplied by chorda tympani fibres in the lingual nerve. Lying behind the vallate papilla on either side is a groove, the **sulcus terminalis.** The two grooves lead back to the foramen caecum. The inferior surface of the free part of the tongue is attached in the midline to the floor of the mouth by a vertical fold, the **frenulum.** Branches of the lingual artery and vein run on each side immediately deep to the mucous membrane towards the tip of the tongue. The sublingual glands form a ridge, the **sublingual fold,** on the floor of the mouth between the tongue and the mandible. The submandibular duct opens on the summit of a small papilla on the side of the frenulum.

The submandibular and sublingual glands (Figs. 16 and 21)

The **submandibular salivary gland** is found lying deep to the middle part of the body of the mandible and the medial pterygoid muscle. The gland has a superficial portion which is covered by the skin and platysma and is crossed by the cervical branch of the facial nerve. The digastric and stylohyoid muscles lie inferior to the more superficial part of the gland. The gland lies on the mylohyoid muscle medially where the facial artery grooves it. The gland extends backwards to the posterior edge of the mylohyoid muscle where it is continuous with its deeper part which lies between the mylohyoid and hyoglossus muscles. The duct comes from this deeper portion and passes forwards medial to the sublingual gland to the papilla on the side of the frenulum of the tongue. The lingual nerve lies on the hyoglossus above the gland but crosses the duct first on its outer (mandibular) aspect and then recrosses it on its deeper (lingual) aspect. The hypoglossal nerve lies between the gland and the hyoglossus and then runs forward on the hyoglossus, inferior to the duct. In front the nerve passes medial to the duct to reach the deeper muscles of the tongue. The gland contains a number of important lymph nodes.

The **sublingual salivary gland** lies below the mucous membrane of the floor of the mouth and has a series of small ducts which open on to the edge of the sublingual fold and into the submandibular duct. (WHARTON'S DUCT)

Both these glands are innervated by postganglionic, secreto-motor branches of the submandibular ganglion (Fig. 23). The preganglionic branches arise in the superior salivary nucleus in the medulla and run in the facial nerve and its chorda tympani branch to the ganglion which is attached to the lingual nerve.

CHAPTER 7

THE DEEP DISSECTION OF THE NECK

INTRODUCTION

MANY of the structures to be dissected and described have already been identified. These will include the thyroid gland which is one of the largest of the endocrine glands and lies mainly on the sides of the trachea.

Examine the cervical vertebrae again and note the anterior and posterior tubercles on the transverse processes. Also identify, on the cadaver and the living subject, in the midline from above downwards, the hyoid bone, and the thyroid, cricoid and tracheal cartilages.

DISSECTION

Clean the structures lying deep and posterior to the sterno-cleidomastoid muscle. Using as a guide the cut ends of the cutaneous nerves that were seen emerging at the posterior border of the sterno-cleidomastoid, define the nerve roots of the cervical plexus from which these branches come. The cervical plexus is formed by the ventral rami of the 1st to the 4th cervical nerves. The **great auricular nerve** comes from the 2nd and 3rd rami. It is not possible at this stage of the dissection to identify the 1st ramus.

Identify the **spinal accessory nerve** as it emerges halfway down the posterior border of the sternocleidomastoid and passes backwards to disappear deep to the trapezius at the junction of its lower one-third and upper two-thirds. The nerve runs along the lower border of the levator scapulae muscle. Inferior to the cervical plexus, find the roots of the brachial plexus, also formed by ventral rami. The rami from the 5th and 6th cervical nerves are seen to join together to form the **upper trunk,** from which the suprascapular nerve arises. The ramus from the 7th cervical nerve forms the **middle trunk** and the rami from the 8th cervical and 1st thoracic nerves join to form the **lower trunk.** Identify the **transverse cervical** and **suprascapular** branches of the subclavian artery as they pass laterally, superficial to the nerves. The ventral rami forming the cervical and brachial

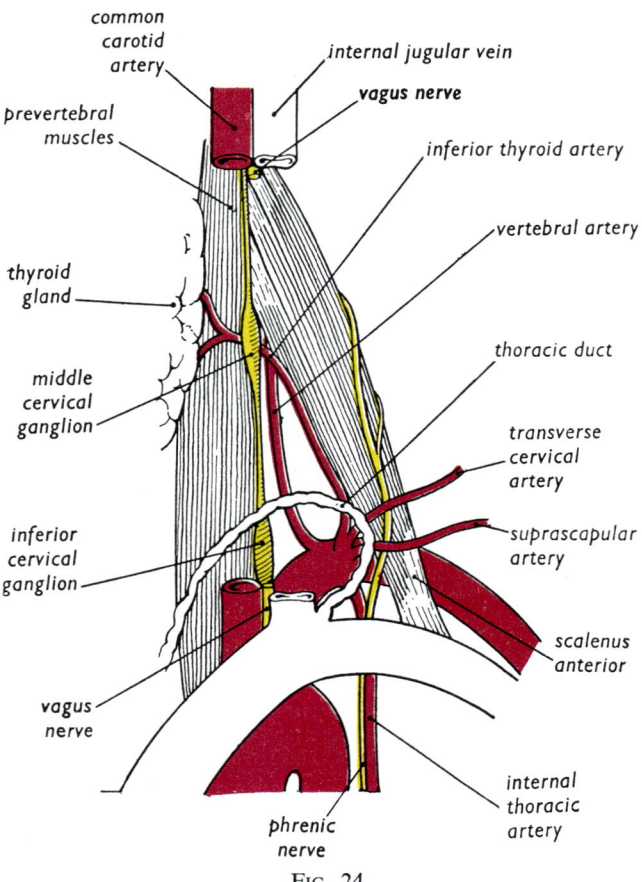

common
carotid
artery

internal jugular vein

vagus nerve

prevertebral
muscles

inferior thyroid artery

vertebral artery

thyroid
gland

thoracic duct

middle
cervical
ganglion

transverse
cervical
artery

inferior
cervical
ganglion

suprascapular
artery

scalenus
anterior

vagus
nerve

internal
thoracic
artery

phrenic
nerve

FIG. 24

The principal relations of the first part of the left subclavian artery. The pleura is not shown.

plexuses pass between the scalenus anterior (attached to the anterior
tubercles of the transverse processes), and the scalenus medius and
levator scapulae (attached to the posterior tubercles). On the
anterior surface of the scalenus anterior find the **phrenic nerve** and
identify its roots (from the 3rd, 4th and 5th cervical spinal nerves).
Remove the clavicle, cutting through the attachments of the sternocleidomastoid and subclavius muscles. Trace the phrenic nerve
into the thorax (Fig. 24).

62

Clean the subclavian vessels and define the branches of the artery. Running upwards and medially are the **vertebral artery** (deep) and the **thyrocervical trunk** (superficial) which divides into the **inferior thyroid artery** (continuing upwards) and the **transverse cervical** and **suprascapular arteries** (both running laterally, superficial to the scalenus anterior muscle and the phrenic nerve). On the surface of the dome of the pleura, find the **costocervical trunk** running backwards over the dome to divide at the neck of the 1st rib into the **superior intercostal artery** (entering the thorax), and the **deep cervical artery** (Fig. 28). The former artery supplies the 1st and 2nd intercostal spaces. Find the **internal thoracic artery** running downwards from the concavity of the subclavian artery.

Find the **sympathetic trunk** lying in the prevertebral fascia behind the carotid sheath (Fig. 25). This nerve has one or two ganglia at the root of the neck and a large **superior cervical ganglion** lying at the level of the angle of the jaw. Fine branches leave the ganglia to run with the spinal nerves and the main blood vessels. Running downwards and medially from each ganglion is a larger **cervical cardiac sympathetic nerve**. The **ansa subclavia** is a loop running between the middle and inferior ganglia and passing round the subclavian artery.

On the left side, find the **thoracic duct** entering the neck from the thorax and lying behind the carotid sheath and in front of the sympathetic nerves. (The right lymph trunk is smaller and more difficult to identify.) The duct passes forwards and laterally. The branches of the subclavian artery lie posterolaterally and the carotid sheath lies anteromedially. The duct joins the junction of the internal jugular and subclavian veins (Fig. 24).

In the floor of the triangle formed by the sternocleidomastoid, trapezius and clavicle find the following muscles from below upwards: the scalenus medius, levator scapulae and splenius capitis (Fig. 12). In the lower half of the triangle, the inferior belly of the omohyoid muscle crosses deep to the trapezius and sternocleidomastoid but superficial to the muscles of the floor.

Clean the infrahyoid group of muscles which lie in the cervical fascia. Superficially are the omohyoid (laterally) and the sternohyoid (medially), and deep to these the thyrohyoid (above) and the sternothyroid (below). The superficial muscles should be

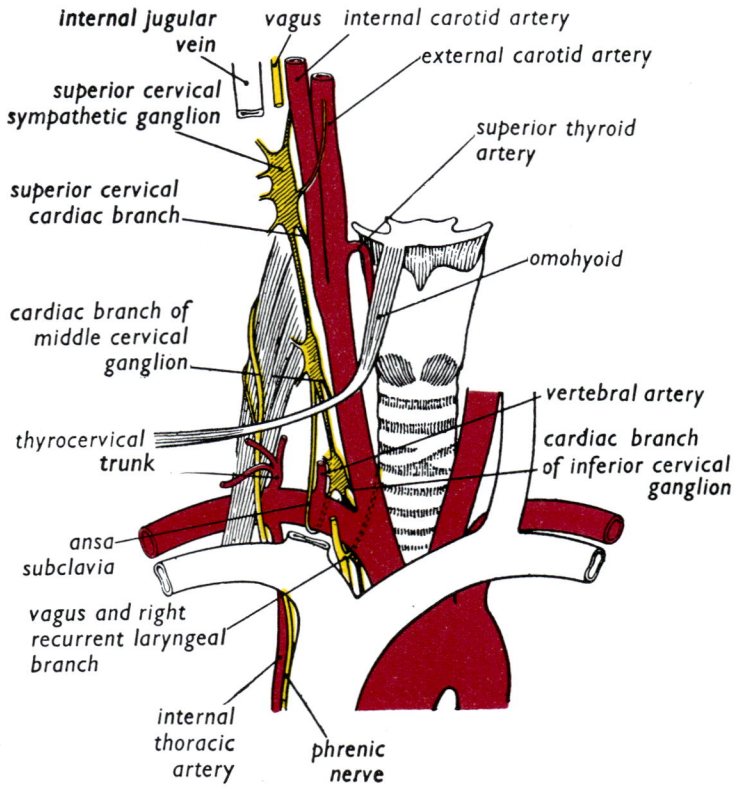

internal jugular vein

vagus   internal carotid artery

external carotid artery

superior cervical sympathetic ganglion

superior thyroid artery

superior cervical cardiac branch

omohyoid

cardiac branch of middle cervical ganglion

vertebral artery

thyrocervical trunk

cardiac branch of inferior cervical ganglion

ansa subclavia

vagus and right recurrent laryngeal branch

internal thoracic artery

phrenic nerve

Fig. 25

The principal relations of the right cervical sympathetic trunk.

detached from the hyoid and turned down to expose the others. Deep to the omohyoid, sternohyoid and sternothyroid muscles, find the thyroid gland (Figs. 14, 15 and 16). Trace the superior and inferior thyroid arteries into the gland and find, with the former, the external laryngeal nerve, and with the latter the recurrent laryngeal nerve. Note that the upper pole of the lateral lobe extends up to the attachment of the sternothyroid muscle to the thyroid cartilage, and the lower pole down to about the sixth ring of the trachea. The isthmus lies opposite the second, third and fourth tracheal rings. There may be a remnant of the thyroglossal duct passing from the isthmus towards the hyoid bone. Detach the gland from the neck taking care not to cut the nerves.

STRUCTURAL DETAILS

The muscles of the neck (Figs. 12, 14, 15 and 16)

The **omohyoid muscle** has two bellies. The superior belly (from the body of the hyoid bone) and the inferior belly (from the upper border of the scapula medial to the suprascapular notch) are joined by a tendon running in a fibrous sheath that is attached to the clavicle and cervical fascia. The **sternohyoid** is attached above to the hyoid bone and below to the posterior surface of the manubrium. The **thyrohyoid** runs between the body of the hyoid and the lamina of the thyroid cartilage above the oblique line. The **sternothyroid** passes from the oblique line on the lamina of the thyroid cartilage to the back of the manubrium.

All these muscles belong to the infrahyoid group. They fix the hyoid bone and therefore are used in swallowing, opening the mouth against resistance, and phonation. They are supplied by the first three cervical ventral rami through the ansa cervicalis, its superior or inferior roots, or the hypoglossal nerve (Fig. 26).

The **scalene muscles** run from the transverse processes of the cervical vertebrae to the 1st rib. The **scalenus anterior** muscle comes from the anterior tubercles of the 3rd to the 6th cervical vertebrae and is attached to the scalene tubercle and upper surface of the 1st rib. The **scalenus medius** muscle comes from the posterior tubercles of the 2nd to the 7th cervical vertebrae and passes to the 1st rib (upper surface), behind the scalenus anterior. A few fibres of the medius are attached to the 2nd rib. Separating the attachments of the anterior and the medius muscles to the 1st rib is a groove in which lie the subclavian artery and a branch of the 1st thoracic nerve. The subclavian vein lies in front of the scalenus anterior. The scalenes, if contracting on both sides, raise the 1st and 2nd ribs and help flex the neck. They may be active in quiet inspiration and are always active in deep inspiration.

The **levator scapulae** runs from the posterior tubercles of the transverse processes of the upper cervical vertebrae to the superior angle of the scapula. It raises the scapula. The **splenius capitis,**

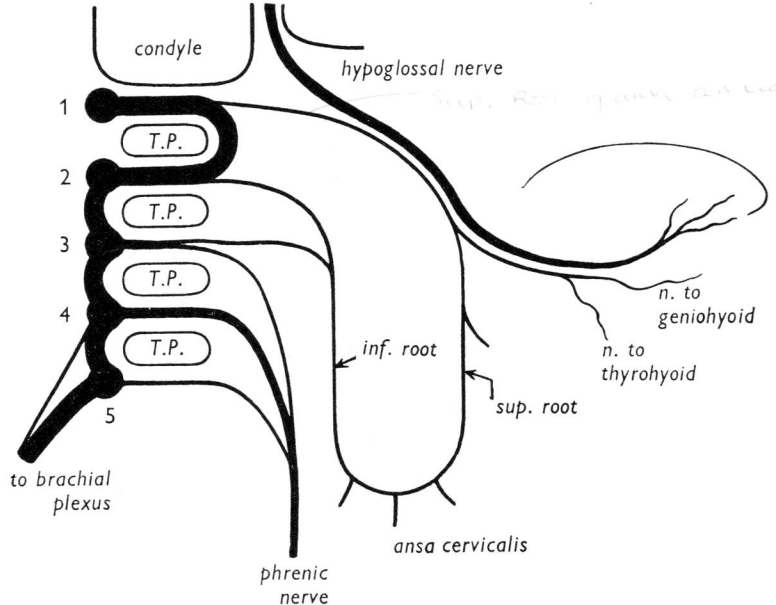

FIG. 26
Diagram showing the formation of the ansa cervicalis.

longissimus capitis and **trapezius** were seen and described in the dissection of the back (pages 8 and 13) and the **sternocleidomastoid** was described on page 41 (Fig. 12).

The nerves in the neck

The **cervical plexus** of nerves is formed by a series of loops joining the ventral rami of the first four or five cervical nerves. The first loop passes in front of the transverse process of the atlas and in this position lies beside the hypoglossal nerve as it emerges from the hypoglossal canal. Fibres from the 1st cervical nerve join the hypoglossal nerve and leave it as the **superior root** of the ansa cervicalis and as branches to the thyrohyoid and geniohyoid muscles. Coming from the 2nd and 3rd nerves are the **inferior root** of the ansa, the **lesser occipital, great auricular** and **transverse cervical nerves.** Other branches are the **supraclavicular** (from the 3rd and 4th nerves), and the **phrenic** (from the 3rd, 4th and 5th

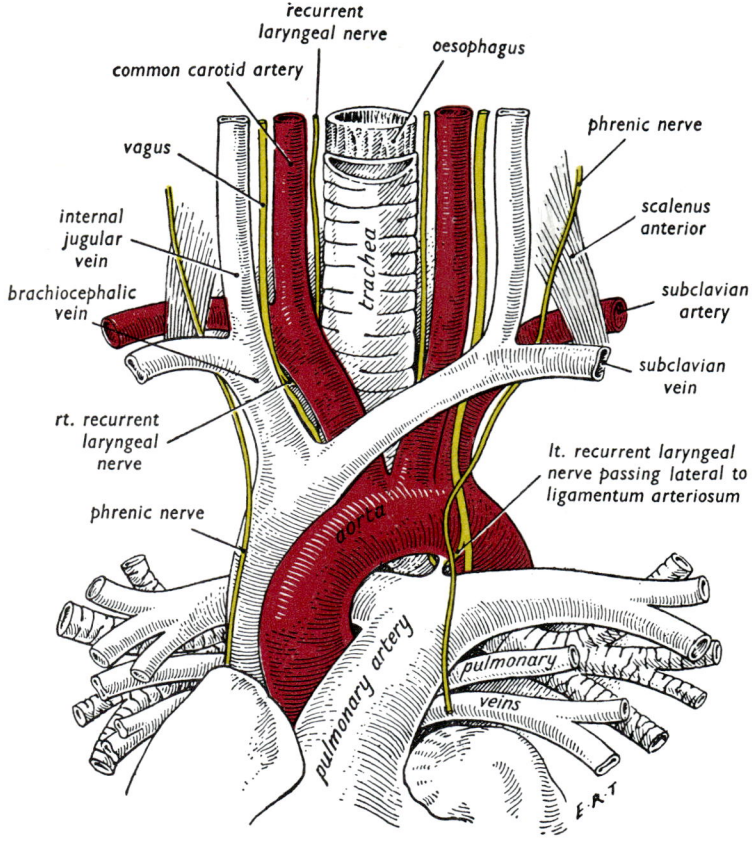

FIG. 27

Diagram of some of the larger vessels and nerves in the upper part of the mediastinum. (The right phrenic nerve has been pulled forwards.)

nerves). Branches are given off to the sternocleidomastoid, trapezius, levator scapulae and scalene muscles.

The **phrenic nerve** arises from the ventral ramus of the 4th cervical nerve and has contributions from the 3rd and 5th nerves (Figs 24 and 27). The branches from these rami turn round the lateral border of scalenus anterior on to the anterior surface of the muscle and form the main nerve which runs down on the muscle as far as the root of the neck. On the scalenus anterior

67

the nerve is crossed anteriorly by the transverse cervical and suprascapular arteries. All these structures lie deep to the sternocleidomastoid muscle. Just before the first rib is reached, the phrenic nerve passes medially leaving the muscle and passes behind the subclavian vein. It lies in front of the subclavian artery on the left side. It crosses in front of the internal thoracic artery and enters the mediastinum where it has already been found and traced to the diaphragm which it supplies. (Revise the thoracic course of the nerve in Vol. I, page 61). Medial to the phrenic nerve is the sympathetic trunk in the prevertebral fascia, and anteromedial are the contents of the carotid sheath.

The **brachial plexus** is formed from the ventral rami of the 5th cervical to the 1st thoracic spinal nerves (Fig. 29). Each ramus is first seen lying between the anterior and posterior tubercles of the transverse processes, i.e. they lie on scalenus medius and behind scalenus anterior. The ventral rami form the **roots** of the brachial plexus. The 5th root gives branches to the rhomboids, the phrenic nerve and to the nerve to serratous anterior and the main part of the nerve joins with the 6th root to form the **upper trunk.** A branch from the 6th root joins the long thoracic nerve and the suprascapular nerve arises from the upper trunk. The 7th root gives a branch to the long thoracic nerve and then continues as the **middle trunk.** The 8th root joins with the 1st thoracic root to form the **lower trunk.** The ventral ramus of the 1st thoracic nerve gives off a small intercostal branch. The main part of the nerve ascends anterior to the neck of the 1st rib, arches forwards over the dome of the pleura and joins with the 8th cervical ventral ramus to form the lower trunk which crosses the first rib and lies behind the subclavian artery (Fig. 28). Branches from the roots supply the scalene and prevertebral muscles.

The **long thoracic nerve** arises mainly from the 6th nerve but receives fibres from the 5th and the 7th nerves. The trunk, having been formed at the lateral edge of the scalenus medius, then runs downwards on the outer surface of the serratus anterior which it supplies.

The **suprascapular nerve** comes off the back of the upper trunk and runs downwards and laterally on the scalenus medius. It follows the inferior belly of the omohyoid to the scapula and passes through the suprascapular notch into the supraspinous fossa. It is accompanied by the suprascapular branch of the subclavian artery and supplies the supraspinatus and infraspinatus muscles.

The three trunks of the brachial plexus each divide into anterior and posterior **divisions**, which in turn form the medial, posterior and lateral **cords**. The divisions lie deep to the clavicle and the cords distal to the clavicle. The cords are named from their relationship to the axillary artery and are described in Vol. III.

The **vagus nerve** leaves the skull along with the glossopharyngeal and accessory nerves through the jugular foramen and has a superior and inferior ganglion just below the foramen. These ganglia contain the cell bodies of the sensory fibres of the vagus. The nerve runs down behind and between the internal jugular vein and the internal and common carotid arteries in the carotid sheath (Figs. 13, 14 and 18). High up in the neck it gives off a **pharyngeal branch,** which passes between the internal and external carotid arteries. Lower down the **superior laryngeal nerve** is given off (Fig. 17) which passes medial to these arteries. Below the hyoid bone this nerve divides into an **internal laryngeal branch** which pierces the thyrohyoid membrane and supplies the mucous membrane of the upper part of the larynx, and an **external laryngeal branch** which runs with the superior thyroid artery and supplies the cricothyroid muscle. Each vagus gives off two **cardiac branches** in the neck. On the **right** side the **recurrent laryngeal branch** of the vagus hooks round the subclavian artery, passing below and then behind the vessel (Fig. 25). The **left recurrent laryngeal nerve** arises in the chest and hooks round the arch of the aorta (Fig. 27).

The **cervical sympathetic trunk** is a continuation into the neck of the thoracic sympathetic trunk (Fig. 25). Most of the fibres arise in the lateral horn of grey matter of the 1st thoracic segment of the cord and synapse in either the inferior, middle, or superior ganglion. The **inferior ganglion** lies just above the neck of the 1st rib posterior to the vertebral artery, is frequently fused with the 1st thoracic ganglion (Figs. 24 and 28) and is often called the **stellate ganglion.** The **middle ganglion** is small and frequently absent. It lies near the inferior ganglion. The **superior ganglion** lies between the hyoid bone and the angle of the mandible on the prevertebral muscles, lateral to the pharynx. The upward continuation of the cervical

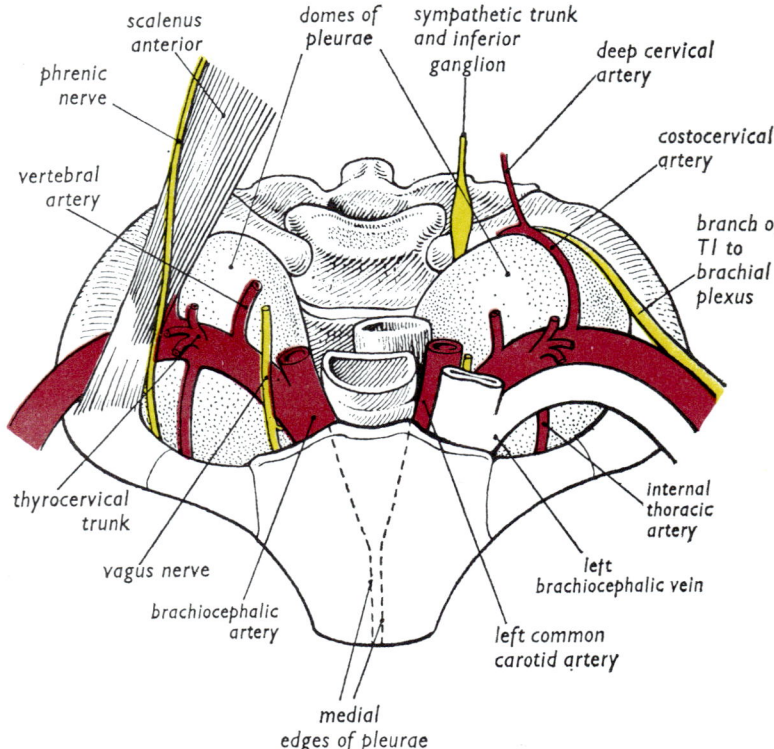

FIG. 28

The structures in the root of the neck. The sympathetic trunk and the veins are not shown on the right and the left scalenus anterior has been removed to expose the left costocervical trunk.

sympathetic trunk is called the **carotid nerve** and accompanies the internal carotid artery into the skull. Postganglionic branches pass from the ganglia to the cervical spinal nerves, to the main blood vessels especially the internal carotid and vertebral arteries, and to the heart (as the superior, middle and inferior **cervical cardiac branches).** In its entire length in the neck, the sympathetic trunk lies on the prevertebral fascia and behind the carotid sheath and its contents. Medial to the sympathetic trunk are the pharynx and larynx above and the oesophagus and trachea below (Fig. 14).

70

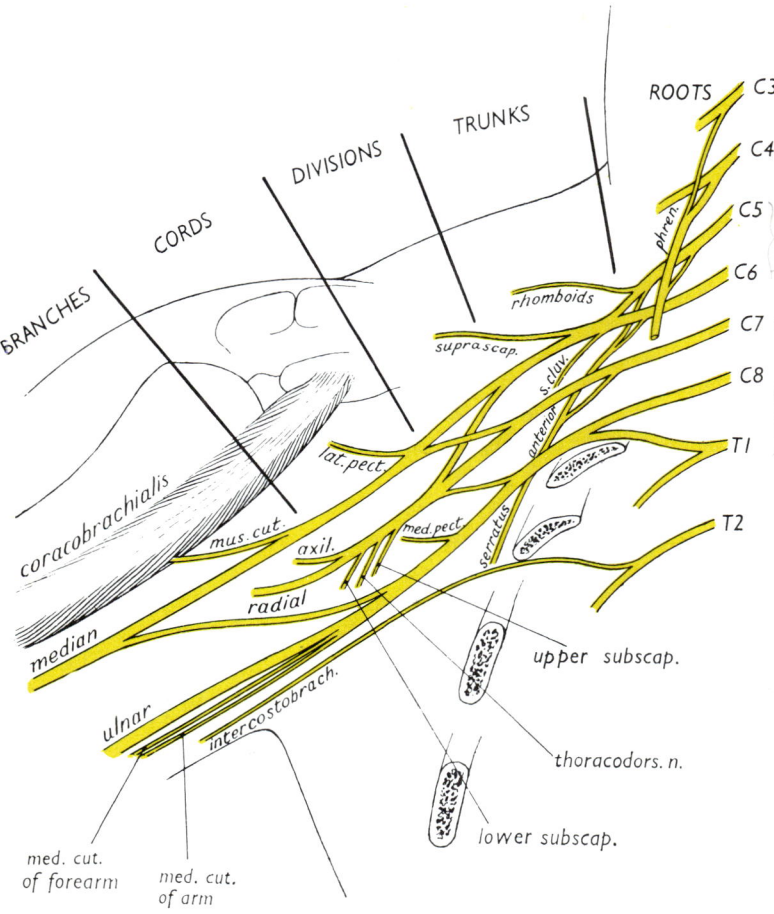

FIG. 29
Diagram of the main features of the brachial plexus.
For description, see text, page 68.

The subclavian vessels (Figs. 25, 27 and 28)

On both sides the **subclavian artery** passes behind the sterno-clavicular joint and then laterally behind the scalenus anterior to cross the first rib and become the axillary artery at the lateral border of the rib. Anterior and inferior to the subclavian artery

71

lies the subclavian vein, separated from it, in its middle third, by the scalenus anterior. Medial to the scalenus anterior the vagus nerve lies in front of the artery. On the left side, the phrenic nerve, anterolateral to the vagus, also lies in front of this part of the artery and the thoracic duct passes forwards across the artery to the brachiocephalic vein. On the right side the vagus nerve gives off its recurrent laryngeal branch which passes upwards and medially behind the artery. The ansa subclavia, sympathetic nerve fibres joining the inferior and middle cervical ganglia, encircles this part of the artery on both sides. The terminal part of the vessel lies behind the clavicle. Posterior to the artery are, medially, the pleura with the apex of the lung and the sympathetic trunk, and laterally, while on the first rib, the lower trunk of the brachial plexus.

The left artery, when in the thorax, is behind the left common carotid artery and left brachiocephalic vein with the left vagus and phrenic nerves intervening. It lies in front of the left border of the oesophagus and the thoracic duct. Medially are the trachea and oesophagus with the left recurrent laryngeal nerve between them, and laterally are the lung and pleura.

Before passing behind the scalenus anterior, the subclavian artery gives off the internal thoracic artery, running downwards into the chest, and the vertebral and thyrocervical arteries, running upwards. (1) The **internal thoracic artery** passes behind the subclavian vein, enters the chest behind the anterior end of the 1st rib and crosses the phrenic nerve (Fig. 24). (2) The **vertebral artery** runs upwards and medially and enters the foramen in the transverse process of the 6th cervical vertebra (Fig. 24). It is accompanied by the vertebral vein. The artery passes in front of the inferior cervical ganglion and is crossed anteriorly by the inferior thyroid artery and the thoracic duct or right lymphatic trunk. The artery ascends through the foramen in the transverse processes of the 6th to the 1st cervical vertebrae, passes medially behind the lateral mass of the atlas and then inferior to the posterior atlanto-occipital membrane (Fig. 5 and page 14). Finally is ascends through the foramen magnum into the skull, pierces the dura and arachnoid mater, and joins the artery of the other side, forming the basilar artery. Inside the skull the vertebral and basilar arteries supply the brain stem, the cerebellum and the occipital part of the

72

forebrain. (3) The **thyrocervical trunk** immediately divides into an inferior thyroid branch going medially and transverse cervical and suprascapular branches going laterally (Fig. 24). The **inferior thyroid artery** passes behind the thoracic duct, in front of the vertebral artery and then behind the sympathetic trunk to reach the inferior pole of the thyroid gland. The **transverse cervical** and **suprascapular arteries** pass behind the thoracic duct and then in front of the phrenic nerve lying on the scalenus anterior. They both lie deep to the omohyoid muscle. (4) Behind the scalenus anterior, the **costo-cervical trunk** is given off (Fig. 28). This vessel passes backwards over the dome of the pleura and at the neck of the 1st rib, divides into the **superior intercostal artery** and the **deep cervical artery**. The former descends in front of the neck of the rib medial to the branch of the ventral ramus of the 1st thoracic nerve going to the brachial plexus, and lateral to the sympathetic trunk, and gives rise to the 1st and 2nd posterior intercostal arteries. The deep cervical branch passes backwards between the transverse processes of the 7th cervical and 1st thoracic vertebrae and ends among the muscles of the back of the neck.

The **subclavian vein** is the continuation of the axillary vein and joins the internal jugular to form the brachiocephalic vein. It lies in front of the scalenus anterior and the phrenic nerve. Its main tributary is the external jugular vein, and, just lateral to the point of entry of the latter, a bicuspid valve is frequently present.

The thoracic duct (Fig. 24)

On the left side the thoracic duct, lying on the prevertebral muscles just behind the oesophagus, passes from the thorax into the neck and crosses anterior to the sympathetic trunk behind the carotid sheath. It runs laterally and forwards in front of the vertebral vessels and the thyrocervical trunk. It sometimes passes as far laterally as scalenus anterior and the phrenic nerve and then turns downwards anterior to the subclavian artery to enter the junction of the subclavian and internal jugular veins. The **right lymph trunk** has a similar course on the right side.

The thyroid gland

The thyroid is an endocrine gland and lies on each side of the larynx and upper part of the trachea (Figs. 14 and 15). It consists of two large **lateral lobes** joined by an **isthmus** which covers the 2nd,

3rd and 4th tracheal rings. The upper pole of the lateral lobe extends up to the oblique line on the thyroid cartilage and the lower pole down to the level of the 6th tracheal ring. The pretracheal fascia forms a fibrous capsule round the gland. On the posterior aspects of the lateral lobes are the four small **parathyroid glands.** In the fresh state, these can often be distinguished from the thyroid but in dissecting room material it is very difficult to find them.

Passing upwards from the isthmus on one or other side of the midline is a small **pyramidal lobe,** which may continue upwards as a fibrous cord, the remnant of the **thyroglossal duct.** The duct develops from the floor of the mouth and, in the adult, its site of origin is marked by the foramen caecum on the dorsum of the tongue.

The arterial supply of the gland is from the superior and inferior thyroid arteries. The former, accompanied by the external laryngeal nerve for part of its course, enters the superficial surface of the upper pole. The inferior thyroid artery passes behind the carotid sheath and, running with the recurrent laryngeal nerve, reaches the deep surface of the gland. The venous blood drains by short vessels directly into the internal jugular vein from the upper parts and into the left or right brachiocephalic veins from the lower parts.

Most of the relations of the gland have been described already except for the medial relations (larynx and pharynx above, trachea and oesophagus below), which will be dissected later. Posteriorly are the prevertebral fascia and muscles and the sympathetic trunk. Superficially are the infrahyoid muscles and laterally the carotid sheath, which is overlapped by the lateral lobe to a certain extent.

The root of the neck

In this region, important structures are found passing between the neck, the thorax and the upper limb (Fig. 28). The upper limb is innervated through the brachial plexus, the proximal part of which (roots and trunks) is found in the neck. The blood supply to the limb leaves the thorax in the subclavian artery and returns in the subclavian vein. The vessels and nerves of the limb pass from the root of the neck into the limb through a space bounded by the clavicle in front, the scapula behind and the 1st rib medially. The dome of the pleura and the apex of the lung project above the anterior part of the 1st rib and form an important posterior relation of the neurovascular bundle of the upper limb. Important structures passing between the thorax and the neck include the large blood vessels, the vagus and phrenic nerves, the sympathetic trunk, and the trachea and oesophagus.

CHAPTER 8

THE MUSCLES OF MASTICATION AND THE TEMPOROMANDIBULAR JOINT

INTRODUCTION

PALPATE on yourself the zygomatic arch formed by a backward projection from the zygomatic bone and a forward projection from the temporal bone. Just below the posterior part of the zygomatic arch and in front of the external acoustic meatus feel the movements of the condylar process of the mandible on opening and closing the mouth. The movements of this process can also be felt by putting a finger into the meatus.

The lateral aspect of the skull is best studied with the mandible disarticulated. A large part of the side of the skull, above and deep to the zygomatic arch, forms the **temporal fossa,** limited below by the **infratemporal crest.** The wall of this fossa, from which the temporalis muscle arises, consists of the frontal and parietal bones above, and the greater wing of the sphenoid and the squamous temporal below. The fossa is limited above by the **superior temporal line** to which the thick fascia covering·the temporalis muscle is attached. The line curves upwards and backwards from the zygomatic process of the frontal bone, across the parietal bone, then downwards and forwards on the squamous temporal bone and ends as the **supramastoid crest** which is continuous with the posterior end of the zygomatic arch. The frontal, parietal, sphenoid and temporal bones meet in an H-shaped suture known as the **pterion,** the centre of which is about 4 cm behind and 2 cm above the frontozygomatic suture. The pterion is of surgical importance as it indicates the position on the inside of the skull where the anterior branches of the middle meningeal vessels lie in a bony groove or canal and are sometimes torn.

Projecting downwards from the body of the sphenoid are the two **pterygoid processes** each consisting of two **pterygoid plates** (medial and lateral). The pterygoid process is separated from the maxilla in front by an opening called the **pterygomaxillary fissure.**

The most posterior part of the alveolar margin of the maxilla is called the **maxillary tuberosity.** The external acoustic meatus is seen behind the posterior end of the zygomatic arch. The **tympanic** part of the temporal bone forms the anterior, inferior and part of the posterior wall of the meatus, whilst the rest of the posterior wall and the superior wall is formed by the **squamous** part of the temporal bone. The **mandibular fossa** on the temporal bone for the head of the mandible lies between the tympanic part and the zygomatic arch. The fossa, in front, is continuous with the **articular tubercle** and behind is limited by the **squamotympanic fissure** lying between the squamous and tympanic parts of the temporal bone. The fissure is subdivided into petrosquamous and petrotympanic parts by a thin projection of the petrous portion of the temporal bone (page 117). The tympanic part of the temporal bone does not form part of the temporomandibular joint and is not covered by cartilage.

DISSECTION

Clean the **masseter muscle** and note its upper attachment to the zygomatic arch. The fibres of the muscle pass downwards (the anterior fibres slightly backwards) to be attached to the lateral surface of the ramus of the mandible.

Above the zygomatic arch the remains of the temporal fascia is seen. Turn it aside by cutting it away from the zygomatic arch, thus exposing the **temporalis muscle,** which spreads out like a fan and is attached to the temporal fascia and to the temporal bone below the superior temporal line. Divide the zygomatic arch with a saw; make the anterior cut well forward but keep the posterior cut in front of the condylar process of the mandible to preserve the temporomandibular joint. The nerve and vessels to the masseter muscle enter the deep surface of the muscle after passing through the mandibular notch. Turn the muscle downwards. The attachment of the temporalis muscle to the coronoid process and to the anterior border of the ramus can now be cleaned. Make an oblique saw cut below the coronoid process from the anterior border of the ramus to the mandibular notch. This cut must be made carefully so as to leave intact the nerve and vessels to the masseter muscle and the buccal nerve that lies deep to the coronoid process. Notice the vertical direction of the anterior fibres of the temporalis muscle and the horizontal direc-

77

tion of its posterior fibres. Turn the coronoid process upwards and clean the deep surface of the muscle. The lateral pterygoid muscle with the masseteric and deep temporal nerves above it and the buccal nerve coming through it will be exposed. The nerves come from the mandibular division of the trigeminal nerve. The first two supply muscles but the third (buccal) is sensory.

A portion of the ramus of the mandible should be removed carefully with a saw and bone forceps. Pass a scalpel handle deep to the ramus of the mandible and separate the bone from the deeper tissues. The inferior alveolar vessels and nerve enter the lower jaw through the mandibular foramen which is on the medial side of the ramus half way between the mandibular notch and the inferior border of the mandible. Make a transverse saw cut across the ramus just above the level of the mandibular foramen. Taking care not to damage the joint or the maxillary artery make another cut through the neck of the condyle of the mandible. Remove the piece of bone and identify the inferior alveolar nerve going into the mandibular foramen, and the lingual nerve anteriorly, both lying on the medial pterygoid muscle. The lateral pterygoid muscle is seen attached to the neck, with the maxillary artery passing either deep to the muscle or superficial to the lower fibres and sinking deeply between its two heads. Passing deep to the lateral pterygoid muscle and the temporomandibular joint and lying close to the inferior alveolar nerve is the **sphenomandibular ligament** passing from the spine of the sphenoid to the lingula. Clean the **lateral pterygoid muscle** and note its two heads, one from the infratemporal fossa and the other from the lateral surface of the lateral pterygoid plate. Trace the muscle to its attachment to the front of the neck of the mandible and to the capsule of the joint. Later it will be seen that it is also attached to the articular disc. The **medial pterygoid muscle** passes downwards and laterally from the medial aspect of the lateral pterygoid plate and from the maxillary tuberosity to the inner aspect of the angle of the mandible (Fig. 30).

Both muscles are supplied by the mandibular division of the trigeminal nerve which leaves the cranial cavity through the foramen ovale in the posterior edge of the greater wing of the sphenoid. The nerve has the small otic ganglion (parasympathetic) on its medial aspect immediately after leaving the skull. These structures

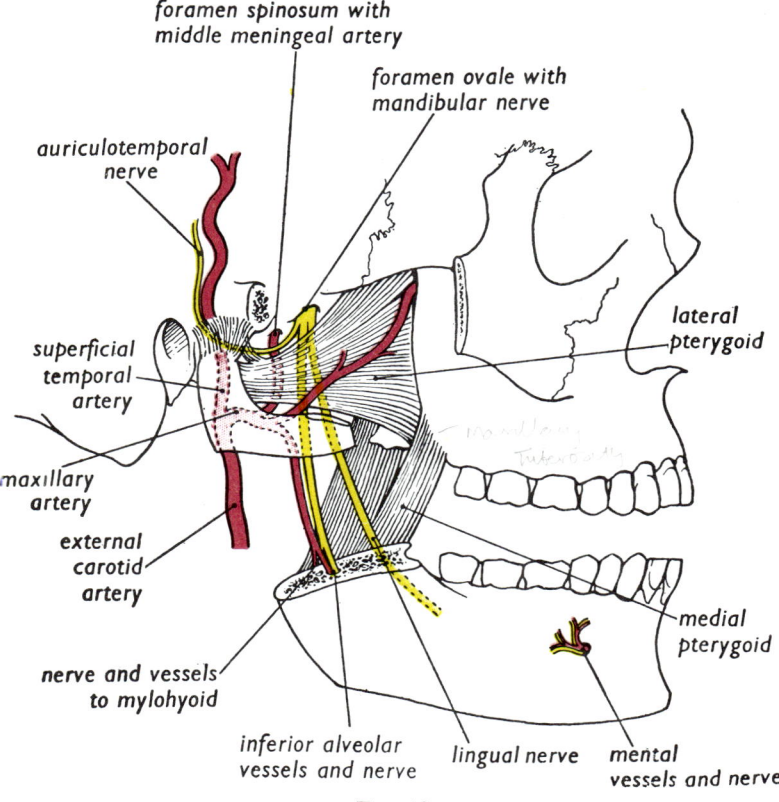

FIG. 30

The pterygoid region. The maxillary artery often passes deep to the inferior head of the lateral pterygoid muscle.

are partly covered in the present dissection by the lateral pterygoid muscle but numerous large branches of the mandibular nerve can now be cleaned and followed. To obtain a better view of some of these nerves reflect the lateral pterygoid muscle forward by cutting its attachment to the mandible.

The **maxillary artery** usually lies superficial to the lateral pterygoid muscle and then enters the pterygopalatine fossa, but it may pass deep to the lower head. It gives off numerous branches as it lies in the pterygoid region and the pterygopalatine fossa. The largest branches are the **middle meningeal** (passing into the skull through

79

the foramen spinosum) and the **inferior alveolar** (entering the mandibular canal). The middle meningeal artery sometimes splits the auriculotemporal nerve soon after the nerve leaves the main mandibular trunk. Branches of the maxillary artery go to all the muscles. Complete the dissection of the branches of the mandibular nerve to the muscles of mastication and the sensory branches (the buccal, passing laterally and then forwards on to the buccinator; the auriculotemporal, passing laterally behind the temporomandibular joint; and the inferior alveolar and lingual, passing downwards between the medial pterygoid muscle and the ramus of the jaw). Near the lower border of the lateral pterygoid muscle find the chorda tympani nerve emerging from the petrotympanic fissure and joining the lingual nerve from behind. The chorda tympani is a branch of the facial nerve.

Examine the temporomandibular joint by cutting the outer aspect of the capsule which is thickened to form the **lateral ligament** passing backwards from the zygomatic arch to the neck of the mandible. The joint is completely divided by an articular disc, which is concave inferiorly to articulate with the condyle and concavoconvex above to articulate with the articular tubercle and mandibular fossa of the temporal bone. Trace the fibres of the lateral pterygoid muscle into the disc.

STRUCTURAL DETAILS

The sphenoid bone

The sphenoid **ossifies** from several centres in the chondrocranium. It has a central **body**, from which two processes extend laterally on each side, the **greater wings** forming part of the middle cranial fossa and of the lateral cranial wall, and the **lesser wings** (above the greater) forming part of the anterior cranial fossa. Two **pterygoid processes** extend downwards each bearing the **medial** and **lateral pterygoid plates.** The lower end of the medial pterygoid plate is elongated to form a hook (the **hamulus**).

The greater wing has a horizontal part in the base of the skull and a vertical part more laterally. The intracranial surface of the horizontal part forms the anterior part of the floor of the middle cranial fossa. Note the **foramen ovale** near the posterior edge and the **foramen spinosum** posterolateral to this foramen. The **foramen rotundum** lies in front of these foramina at the side of the body of the sphenoid bone. The extracranial surface of the horizontal part forms most of the infratemporal fossa and the bony surface lying lateral to the upper part of the lateral pterygoid plate. Note the foramen ovale and the foramen spinosum with the **spine** of the sphenoid posterolaterally. The outer surface of the vertical part of the greater wing forms a part of the wall of the temporal fossa, the bony surface situated medial and superior to the zygomatic arch. These infratemporal and temporal surfaces are separated by the sharp **infratemporal crest.** The anteromedial surface of the greater wing forms the greater part of the lateral wall of the orbit.

The lesser wings extend laterally on each side of the anterior extremity of the body of the sphenoid and form the posterior part of the floor of the anterior cranial fossa. Anteriorly they articulate with the orbital plates of the frontal bone ; posteriorly they present free concave borders which end medially as the anterior clinoid processes and are separated from the greater wings by the **superior orbital fissures.**

The pterygoid process extends downwards from the junction of the body with the greater wing. The anterior borders of the plates are joined together above but are separated below ; the posterior borders are widely separated.

The mandible and the temporomandibular joint

The mandible has already been described on pages 27 and 51. Note the coronoid process for the attachment of the temporalis muscle and the hollow in front of the neck for the lateral pterygoid muscle. The condyles are elliptical in shape, the long axes being directed posteromedially.

The temporomandibular joint is a synovial joint in which an articular disc separates the condyle of the mandible from the temporal bone. The capsule of the joint is thickened laterally to form the strong **lateral ligament.** Other ligaments assist the muscles in attaching the jaw to the skull. The most important of these is the **sphenomandibular ligament,** which passes between the spine of the sphenoid and the lingula of the mandible. The **stylomandibular ligament** joins the tip of the styloid process to the posterior border of the ramus above the angle and is a condensation of the cervical fascia. The **pterygomandibular raphe** joins the hamulus of the medial pterygoid plate to the posterior end of the mylohyoid line.

The pterygoid muscles

The **lateral pterygoid muscle** has an upper head arising from the infratemporal surface of the greater wing of the sphenoid and a lower head from the lateral surface of the lateral pterygoid plate. The fibres of the muscle converge as they pass backward to be attached to the front of the neck of the mandible, and to the capsule and the articular disc of the temporomandibular joint. The **medial pterygoid muscle** is attached to the medial aspect of the lateral pterygoid plate. Some of the muscle fibres rise from the tuberosity of the maxilla. The medial pterygoid passes downwards and laterally to be attached to the medial surface of the angle and ramus of the mandible.

The mandibular division of the trigeminal nerve (Fig. 30)

The mandibular nerve leaves the skull through the foramen ovale and immediately divides into a number of branches.

(1) The **inferior alveolar nerve** lies between the medial and the lateral pterygoid muscles. Inferior to the lower border of the latter muscle, it lies between the medial pterygoid and the mandible and enters the mandibular canal at the mandibular foramen. Before the nerve enters the mandibular foramen the **mylohyoid branch** is given off. It keeps close to the medial aspect of the mandible below the mylohyoid line and supplies the mylohyoid and the anterior belly of the digastric muscle. Anteriorly, at the mental foramen, the inferior alveolar nerve gives off the **mental branch** to the skin and mucous membrane of the lower lip and chin. The right and left inferior alveolar nerves supply all the teeth of the lower jaw, the central incisors being supplied by both nerves.

(2) The **lingual nerve** is anterior to the inferior alveolar nerve and at first has the same relations. It leaves the space between the medial pterygoid muscle and the mandible and lies against the bone below and behind the 3rd molar tooth, being covered only by mucous membrane. It passes on to the hyoglossus where it has been dissected (page 58). The lingual nerve passes forwards on the side of the tongue to supply the anterior two-thirds of its mucous mem-

brane. Just at or near the inferior border of the lateral pterygoid muscle the **chorda tympani** branch of the facial nerve joins the lingual nerve from behind. The chorda tympani emerges from the

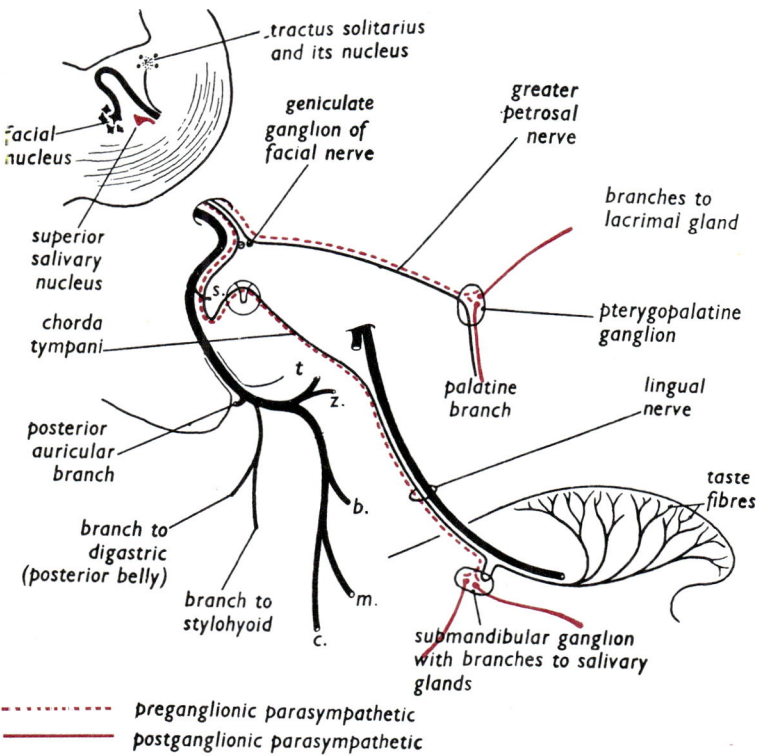

FIG. 31

Diagram of origin, course and constituents of the facial nerve. The main part of the nerve supplies the muscles derived from the second pharyngeal arch (s, stapedial; t, temporal; z, zygomatic; b, buccal; m, mandibular; and c, cervical branches). The sensory fibres (taste) and the parasympathetic secretomotor fibres are shown. Inset is a section of the pons showing the positions of the constituent nuclei.

skull through the petrotympanic fissure and carries taste fibres from the anterior two-thirds of the tongue and preganglionic secretomotor fibres (parasympathetic) to the submandibular and sublingual salivary glands (Figs. 23 and 31).

(3) The **buccal nerve** lies deep to the coronoid process of the mandible and then on the buccinator muscle. It supplies sensory fibres to the skin and mucous membrane of the cheek.

(4) The **auriculotemporal nerve** at first lies medial to the condyloid process of the mandible and sometimes encircles the middle meningeal artery. The nerve passes laterally behind the neck of the mandible and then upwards behind the joint with the superficial temporal artery. It carries sensory fibres from the tragus (the projection anterior to the external acoustic meatus), the lateral surface of the auricle, the anterior surface of the meatus, and the temporal region of the scalp. The auriculotemporal nerve also carries the postganglionic secretomotor fibres (parasympathetic) to the parotid gland. These come from nerve cells in the otic ganglion (Fig. 32).

(5) The motor branches of the mandibular division of the trigeminal nerve, in addition to the **mylohyoid nerve** already described, are (i) the **nerve** to the **masseter** passing laterally through the mandibular notch to the masseter, (ii) the **nerves** to the **medial** and **lateral pterygoid** muscles which enter the deep surfaces of these muscles and (iii) the **deep temporal nerves** passing laterally above the superior border of the lateral pterygoid muscle to enter the deep surface of the temporalis muscle. (iv) Small twigs from the nerve to the medial pterygoid pass through the otic ganglion without synapsing, and supply the **tensor tympani** and the **tensor veli palatini** muscles.

The maxillary and superficial temporal arteries (Figs. 11 and 30)

The external carotid artery divides behind the neck of the mandible into the superficial temporal artery and the maxillary artery. the **maxillary artery** is the larger branch and passes forwards between the neck of the mandible and the sphenomandibular ligament. It usually runs superficial to the lower head of the lateral pterygoid muscle and then medially between the two heads. It may lie deep to both heads before being lost to view as it passes into the pterygopalatine fossa. In this short course it gives off a large number of branches: (1) to the **ear** and **tympanic cavity**, (2) the **middle meningeal artery** which ascends deep to the lateral pterygoid muscle and enters the skull through the

84

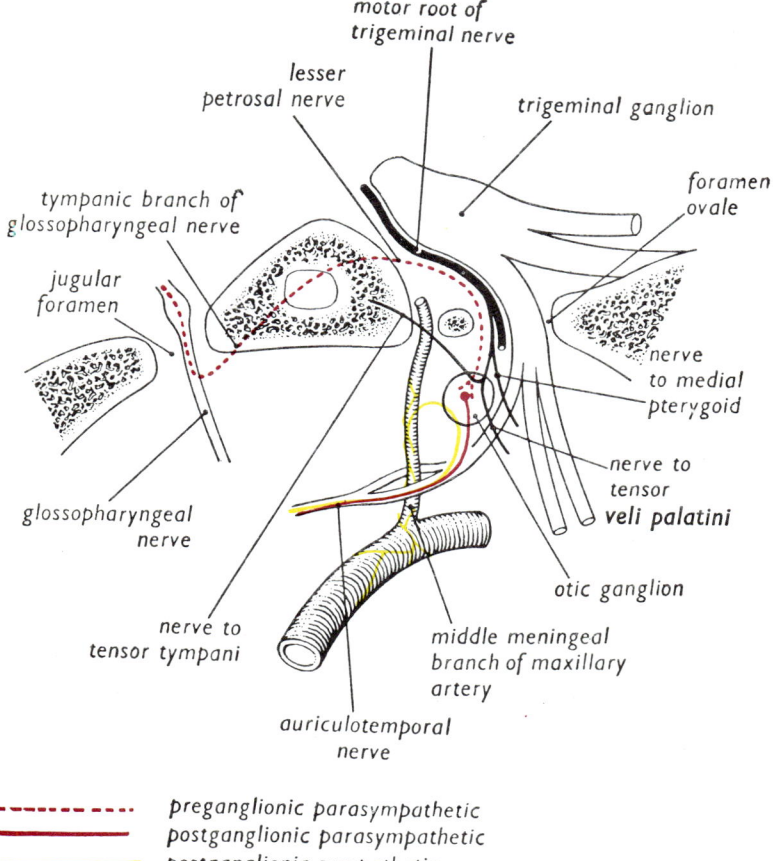

motor root of
trigeminal nerve

lesser
petrosal nerve

trigeminal ganglion

tympanic branch of
glossopharyngeal nerve

foramen
ovale

jugular
foramen

nerve
to medial
pterygoid

glossopharyngeal
nerve

nerve to
tensor
veli palatini

nerve to
tensor tympani

otic ganglion

middle meningeal
branch of maxillary
artery

auriculotemporal
nerve

- - - - - - - - - - preganglionic parasympathetic
———————— postganglionic parasympathetic
———————— postganglionic sympathetic

Fig. 32

Diagram of the position and connexions of the otic ganglion. The postganglionic parasympathetic fibres pass in the auriculotemporal nerve to
the parotid gland.

foramen spinosum, (3) the **inferior alveolar artery** which lies
posterior to the inferior alveolar nerve and is distributed with the
nerve, (4) to the neighbouring **muscles,** *i.e.* masseter, pterygoids,
buccinator, temporalis, (5) numerous branches in the pterygopalatine fossa to the **nose** and **palate** (page 153). The **superficial**

85

temporal artery lies on the posterior end of the zygomatic arch anterior to the auriculotemporal nerve and passes upwards into the scalp. Its branches ramify over the side of the forehead and temporal region. During life pulsation of the artery can frequently be seen.

The **pterygoid plexus** of veins is met throughout the dissection and drains by way of the maxillary vein to the retromandibular vein. The plexus also communicates with the veins of the orbit and with the cavernous sinus inside the skull.

FUNCTIONAL ASPECTS

The movements of the mandible

When the muscles attached to the mandible are relaxed, as in anaesthesia or sleep, the mouth opens under the influence of gravity. Hinge and gliding movements occur at the joints, usually in combination, from the position of rest, in which the teeth are slightly separated.

A. **Hinge Movements.** The hinge movements occur between the condyles of the mandible and the articular discs and are brought about by the masseter, temporalis and medial pterygoid muscles working against gravity and the hyoid group of muscles working with gravity.

B. **Gliding.** In gliding, the condyles of the mandible and the articular discs are pulled forwards by the lateral pterygoids on to the articular tubercles. Restoration to the original position is brought about by relaxation of the lateral pterygoids and by contraction of the posterior fibres of both temporalis muscles.

(1) Closure of the mouth and forcible approximation of the teeth are associated with elevation of the anterior part of the mandible. They are produced by the masseter, temporalis and medial pterygoid muscles. (2) In opening the mouth (depression of the anterior part of the mandible) the lateral pterygoid muscles pull the condyles and discs forwards on to the articular tubercles and the weight of the jaw opens the mouth. The mylohyoid, digastric and geniohyoid muscles depress the jaw against resistance. (3) Side to side movements are produced by alternate contraction of the pterygoid muscles of each side. The right lateral and medial pterygoid muscles rotate the jaw to the left at the left temporomandibular

joint. (4) In chewing food the mandible moves slightly from side to side and is at the same time elevated by the masseter and temporalis muscles.

The buccinator and the tongue help the muscles of mastication in that they push the food back between the teeth.

CHAPTER 9

THE INTRACRANIAL REGION

INTRODUCTION

THE large cavity inside the skull is filled by the brain and its membranes containing the cerebrospinal fluid. The outer membrane (dura) is firmly attached to the bones over the base of the skull but can be stripped more easily from the vault. There is no extradural space comparable with that found in the vertebral canal. Over most of its extent the dura consists of two closely connected layers but in certain regions the layers separate to surround venous sinuses that drain the blood from the brain. The innermost layer of dura forms two prominent folds. One, the **falx cerebri,** lies in the median sagittal plane and partially separates the two cerebral hemispheres from each other. The other, the **tentorium cerebelli,** is in an oblique plane and supports the posterior part of the cerebrum, separating it from the cerebellum. The tentorium is attached to the occipital and temporal bones and has an anterior free edge which lies close to the union of the forebrain with the midbrain. The capacity of the cranial cavity increases with the growth of the brain. In the newborn infant, there are large spaces between the individual bones (fontanelles) and the edges of the bone are not yet serrated. In the adult, the bones are firmly interlocked at the sutures and the skull forms an excellent protection for the brain. Revise Chapter 1 where the bones of the skull are identified (Fig. 1).

DISSECTION

Separate the scalp from the skull so that the two sides can be pulled down over the ears, eyes and occiput. Attach the metal " coronet " firmly to the skull just above the supra-orbital ridges, the external acoustic meatus and the external occipital protuberance and saw through the vault. The final part of the removal of the calvarium is best done with a chisel so that damage to the underlying dura mater is minimal. Strip the dura from the inner surface of the calvarium and make a sagittal midline incision

88

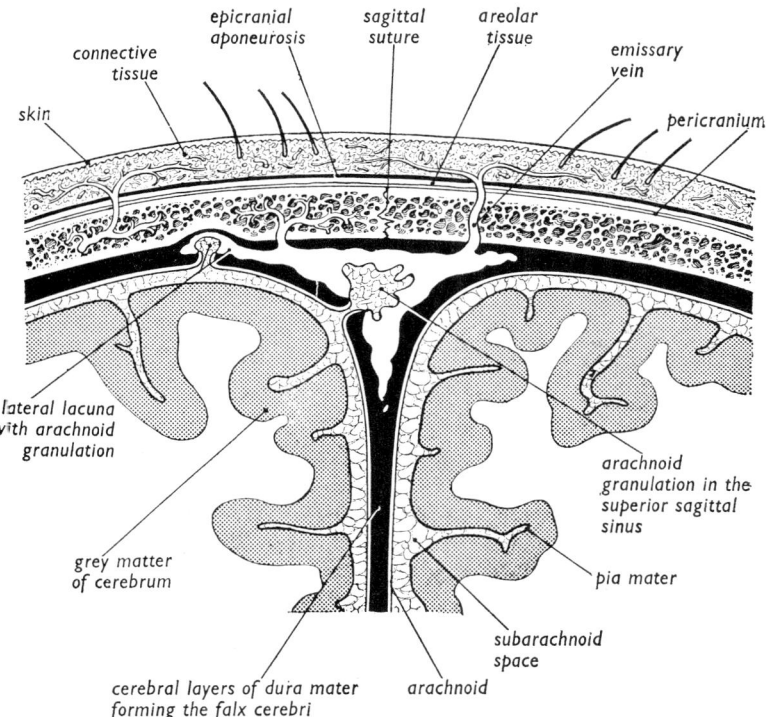

FIG. 33

Diagram of a coronal section across the midline of the vault of the skull.

in the dura from front to back, thus opening the **superior sagittal venous sinus** which extends backwards in the upper edge of the falx cerebri from the crista galli of the ethmoid bone in front to the internal occipital protuberance. Remove the blood clot and note the fibrous bands within the sinus and the **lateral recesses** or **lacunae** bulging outwards on either side of the main sinus. These lacunae produce a series of round depressions in the calvarium along the sides of the groove for the sinus. Open one of the lacunae and note that it is filled by a glomerular structure, the **arachnoid granulation** which transfers the cerebrospinal fluid from the subarachnoid space into the venous sinuses (Fig. 33).

Incise the dura round the cut edge of the skull and cut the

89

attachment of the falx cerebri to the crista galli. The dural cap can now be turned backwards and the falx cerebri examined. In its lower free edge is the **inferior sagittal sinus,** passing backwards from the crista galli to the **straight sinus** which lies in the attachment of the falx cerebri to the tentorium cerebelli. Retract the frontal lobe of the cerebrum and between it and the base of the skull, identify the **olfactory tract** running back from the **olfactory bulb** which lies on the upper surface of the **cribriform plate** on either side of the crista galli. The **optic nerve** is large and can be traced back to its union with the nerve of the opposite side in the **optic chiasma.** The 3rd **(oculomotor)** and 4th **(trochlear)** cranial nerves lie near the free edge of the tentorium cerebelli. If the brain is firm, lift up the frontal lobes and cut through the olfactory and optic nerves, the internal carotid arteries, and the brain-stem at the level of the tentorium, so that the forebrain can be removed in one piece. The frontal lobes of the forebrain lie in the **anterior fossa** of the skull and the temporal lobes in the **middle fossa.** The occipital lobes lie on the tentorium cerebelli which forms the roof of the **posterior cranial fossa.** The posterior fossa contains the hindbrain which includes the cerebellum. In the midline behind the optic chiasma the narrow **infundibulum** will be seen passing down to the hypophysis cerebri (pituitary gland).

Examine the tentorium (Fig. 34). Its outer margin is attached to the occipital bone and the petrous part of the temporal bone and forms a circle, deficient in front. From this attachment, it rises like a tent but leaves a circular opening in front formed by the free margin of the tentorium and the body of the sphenoid. Through the opening passes the midbrain. Open the straight sinus and trace it and the superior sagittal sinus to the internal occipital protuberance where they continue into the **transverse sinuses.** Usually the superior sagittal turns to the right as the right transverse sinus and the straight to the left as the left transverse sinus. The transverse become the **sigmoid sinuses.**

Make incisions in the dura on either side of the straight sinus and along the transverse sinus and turn the tentorium forwards, thus exposing the hindbrain. The 5th cranial nerve **(trigeminal)** leaves the pons and passes forwards below the anterior end of the attached border of the tentorium. The 6th **(abducent)** is a small nerve which pierces the dura below and medial to the trigeminal nerve. Both the above nerves, with the 3rd and 4th, pass forwards but the 7th **(facial)** and 8th **(vestibulocochlear)** pass laterally and enter the internal acoustic meatus which lies on the posterior surface

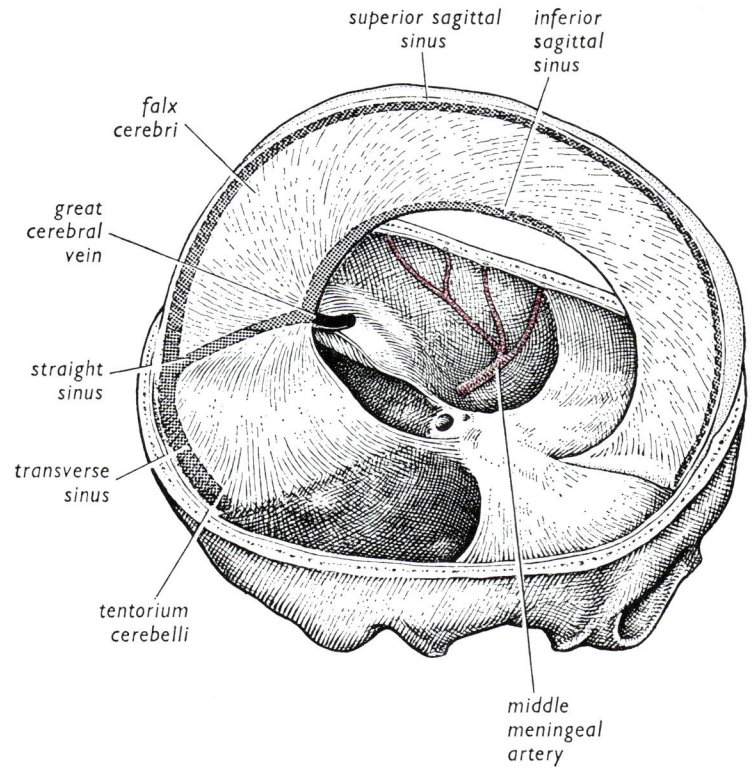

FIG. 34

The vault of the skull and much of its dural lining has been removed to expose the falx cerebri and the tentorium cerebelli.

of the petrous temporal posterolateral to the trigeminal nerve. The 9th **(glossopharyngeal)**, 10th **(vagus)** and 11th **(accessory)** nerves pass downwards and laterally and pierce the dura at the **jugular foramen** which also transmits the inferior petrosal and sigmoid venous sinuses. The 12th **(hypoglossal)** nerve passes laterally out of the skull through the **hypoglossal canal** very near the edge of the large centrally placed **foramen magnum** which transmits the brain stem, vertebral arteries and spinal accessory nerves. This nerve can be followed laterally to the jugular foramen where it joins the cranial accessory and leaves the skull. Cut through the cranial nerves leaving the brain-stem and cut across the brain-stem and vertebral vessels in the foramen magnum. Remove the hindbrain and expose the posterior cranial fossa. Open the transverse sinuses and trace them forwards and downwards to the jugular foramina. The curved portion anteriorly is referred to as the **sigmoid sinus.**

The **hypophyseal fossa** is roofed over by a layer of dura, called the **diaphragma sellae** which is pierced about its middle by the infundibulum. The diaphragma is attached to four bony promi-nences, the **anterior** and **posterior clinoid processes.** Remove the diaphragma and define the positions of the clinoid processes. The posterior clinoid processes are part of the dorsum sellae, the back wall of the hypophyseal fossa. The 3rd and 4th cranial nerves are anterior to the posterior clinoid processes and the 6th nerve is posterior.

Lateral to the posterior clinoid process in the middle cranial fossa find the opening of the pocket of dura mater containing the trigeminal ganglion. Open this pocket by cutting through its roof, expose the ganglion and trace its divisions forwards (Fig. 35). The **ophthalmic** (medial) and **maxillary** (middle) divisions both run forwards and the **mandibular** division, the most lateral, passes downwards to the foramen ovale. Find the motor root passing deep to the ganglion and joining the mandibular nerve. Carefully remove the dura from the side of the body of the sphenoid, thus opening the **cavernous sinus.** It is best to trace the oculomotor, trochlear and abducent nerves forwards into the sinus and remove the dura of the lateral wall as the dissection proceeds.

Note that the maxillary nerve has only a short course in the cavernous sinus and then passes through the **foramen rotundum.** The ophthalmic division runs forwards in the lateral wall of the sinus below the oculomotor and trochlear nerves and divides at its anterior end into **lacrimal, frontal** and **nasociliary** branches (Fig. 36). The abducent nerve is inferior and then lateral to the **internal carotid artery** which, having passed through the petrous portion of the temporal bone, passes medially across the foramen lacerum and turns forwards with the abducent nerve in the cavernous sinus. The internal carotid artery turns upwards medial to the anterior clinoid process, pierces the dura and was cut here when the brain was removed. It is closely related to the optic nerve and chiasma and gives off ophthalmic and hypophyseal branches, as well as small twigs to the meninges.

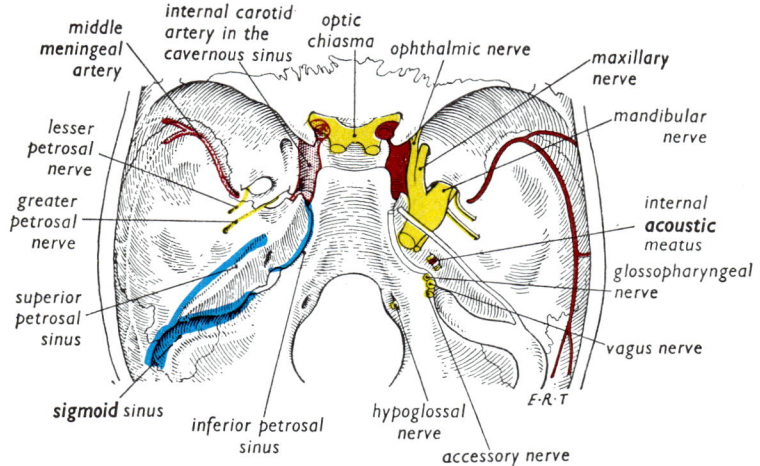

FIG. 35

The inside of the base of the skull, indicating some of the relations of the trigeminal ganglion.

STRUCTURAL DETAILS

The internal aspect of the vault

The following markings should be observed on the bones.

1. The sagittal groove, for the superior sagittal (venous) sinus, passes backwards in the midline to the internal occipital protuberance, where it becomes continuous with the groove for the transverse sinus of one side (more commonly the right).

2. Large ill-defined depressions produced by the lateral lacunae. (These are lateral expansions of the superior sagittal sinus and small pits in the depressions are made by the arachnoid granulations.)

3. The groove for the middle meningeal vessels passes laterally from the foramen spinosum.

93

4. Shallow impressions produced by the cerebral gyri.
5. The parietal foramina, in the region of the vertex, for the passage of emissary veins which link up the intracranial venous sinuses and extracranial scalp veins.

The cranial fossae

A. The **anterior cranial fossa** contains the frontal lobes of the brain. The greater part of the floor is formed by the **orbital plates** of the frontal bone, and separates the frontal lobes of the brain from the orbits. The fossa is separated from the nasal cavity by the **cribriform plates** of the ethmoid, so called because of the perforations in them for the passage of bundles of olfactory nerves. The crista galli projects upwards in the midline and provides attachment for the anterior end of the falx cerebri. The posterior part of the floor of the fossa is formed by part of the body of the sphenoid in the midline and by its lesser wings laterally. The free border of a lesser wing when traced medially ends in an anterior clinoid process.

The close relationship of the anterior cranial fossa to the roof of the nose and of the orbit is important clinically. Fracture of the cribriform plate may tear the meninges and allow the cerebrospinal fluid to escape. Infection may also spread upwards from the nose.

B. The **middle cranial fossa** is divided into a small median and two larger lateral parts.

The median part is formed by the body of the sphenoid, and has in front a shallow groove and behind a much deeper one, the **sella turcica** or **hypophyseal (pituitary) fossa.** The shallow groove traced laterally leads to the **optic canal** which opens into the back of the orbit. The optic chiasma is not situated in this groove, but lies behind it. The sella turcica is separated from the posterior cranial fossa by the **dorsum sellae.** The corners of the dorsum sellae form the posterior clinoid processes.

On the lateral surface of the body of the sphenoid is the **cavernous venous sinus** and, passing forwards in the sinus, are the internal carotid artery (frequently grooving the bone) and several cranial nerves. In the body of the sphenoid are the right and left **sphenoidal air sinuses** which communicate with the nasal cavity.

The lateral part of the middle fossa lodges the temporal lobe of the brain, and is limited in front by the posterior margin of the lesser wing of the sphenoid and behind by the obliquely placed

94

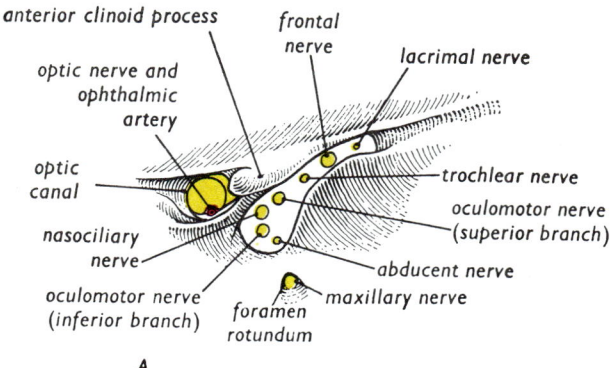

A

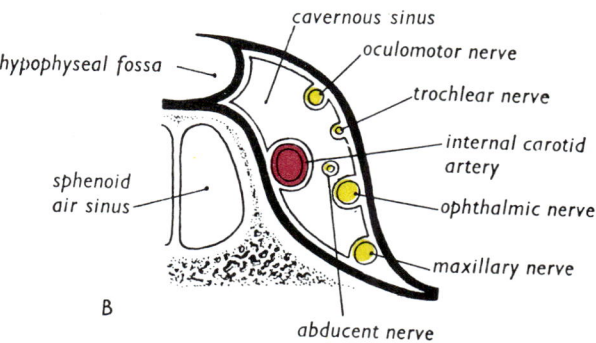

B

FIG. 36

A. Some of the structures related to the right optic canal and the right superior orbital fissure. B. The posterior end of the right cavernous sinus, coronal section.

superior border of the petrous part of the temporal bone. Its floor is formed by the greater wing of the sphenoid in front, the anterior surface of the petrous temporal bone behind and the squamous part of the temporal bone laterally.

Related to the greater wing of the sphenoid are five openings (Fig. 35).

1. The overhanging lesser wing is separated from the greater wing by a gap, the **superior orbital fissure,** which leads into the orbit and transmits nerves and vessels (Fig. 36A and page 101).

95

2. Immediately behind the medial end of this fissure is the **foramen rotundum** which transmits the maxillary nerve and opens anteriorly into the pterygopalatine fossa. Confirm this on a skull by means of a probe.

3. The **foramen ovale** is in the greater wing of the sphenoid about 2 cm behind the foramen rotundum. It leads directly downwards into the infratemporal fossa and transmits the mandibular nerve.

4. The **foramen spinosum** lies close to the posterolateral border of the foramen ovale and transmits the middle meningeal vessels from the infratemporal fossa. The vessels groove the bone as they pass laterally and are liable to be torn following injury to the skull.

5. The **foramen lacerum** (Fig. 35) is a large irregular opening medial to the foramen ovale and anterior to the apex of the petrous temporal. The foramen is also lateral to the occipital bone and the body of the sphenoid when seen from below. In life the lower part of the foramen lacerum is filled by fibrocartilage. The internal carotid artery enters the skull by the **carotid canal** which runs in the petrous temporal bone and emerges from the posterolateral wall of the foramen lacerum to lie above the cartilage. The artery then passes forwards in the carotid groove at the side of the body of the sphenoid and turns upwards medial to the anterior clinoid process. The **pterygoid canal** passes forwards in the sphenoid from the anterior aspect of the foramen lacerum above the cartilage to the pterygopalatine fossa and transmits the **nerve of the pterygoid canal** formed by the union of the greater petrosal (parasympathetic) and the deep petrosal (sympathetic) nerves.

The anterior surface of the petrous temporal bone has lateral to its apex a shallow depression for the **trigeminal ganglion.** A smooth elevation, the **arcuate eminence** is also seen on the anterior surface of the petrous temporal near its upper edge. It is produced by the convexity of the **anterior semicircular canal** of the internal ear which lies in the substance of the bone. Lateral to the arcuate eminence a thin plate of bone, the tegmen tympani, forms the roof of the middle ear cavity and the tympanic antrum. The superior border of the petrous temporal is grooved by the superior petrosal venous sinus which passes from the cavernous to the transverse sinus. Infection in the ear cavities may spread through the bone to involve the meninges, sinuses and brain.

96

C. The **posterior cranial fossa** lodges the hindbrain which consists of the pons, cerebellum and medulla oblongata. The fossa is bounded in front by the dorsum sellae, anterolaterally by the superior border of the petrous temporal bone, and posteriorly by the margins of the shallow groove on the parietal and occipital bones produced by the transverse venous sinuses, and in the midline by the internal occipital protuberance.

In the floor of the fossa is the **foramen magnum.** In front of this the floor slopes upwards, being formed by the basilar part of the occipital bone and the body of the sphenoid bone. The medulla and the pons with the basilar artery lie on this part of the skull. The remainder of the floor is formed anterolaterally by the petro-mastoid part of the temporal bone and posteriorly by the condylar and squamous parts of the occipital bone. The **occipital condyles** are seen encroaching on the anterior part of the foramen magnum. Above and anterior to each condyle is the opening of the **hypoglossal canal** which passes laterally and forwards and transmits the hypoglossal nerve.

The groove for the transverse venous sinus begins at the internal occipital protuberance and at first passes horizontally on the occipital and parietal bones to the lateral extremity of the superior border of the petrous temporal bone, below which there is an S-shaped groove on the temporal and occipital bones for the sigmoid sinus. Two foramina for emissary veins may be found opening into or near the groove for the sigmoid sinus, the **condylar foramen,** immediately behind the occipital condyle, and the **mastoid foramen** passing through the mastoid part of the temporal bone.

The **jugular foramen** is bounded in front by the petrous temporal and behind by the occipital bone. The petro-occipital fissure in front of the jugular foramen is grooved by the **inferior petrosal venous sinus.** This begins at the posterior end of the cavernous sinus and ends in the internal jugular vein after passing through the anterior part of the jugular foramen (Fig. 37). The sigmoid sinus becomes continuous with the internal jugular vein after passing through the posterior part of the foramen. The glosso-pharyngeal, vagus and accessory nerves pass through the middle part of the foramen.

97

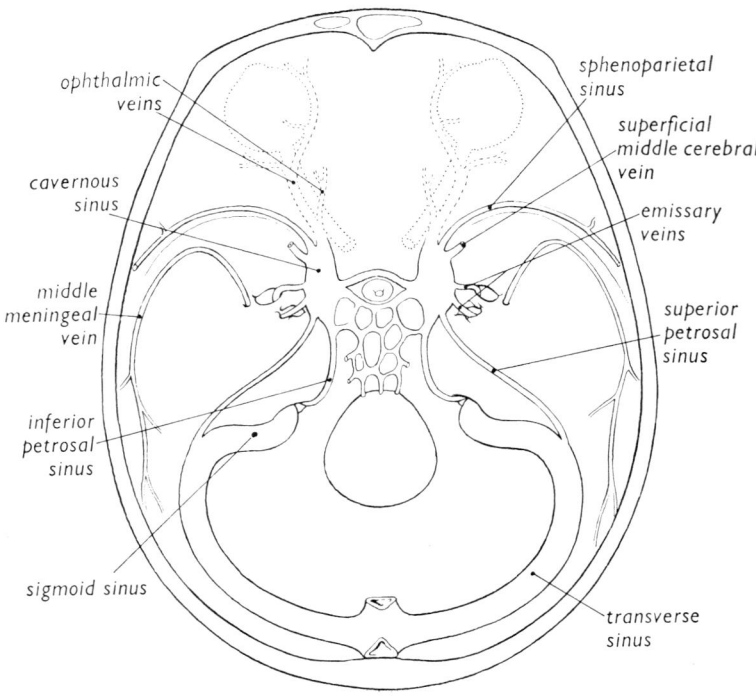

ophthalmic veins

sphenoparietal sinus

cavernous sinus

superficial middle cerebral vein

emissary veins

middle meningeal vein

superior petrosal sinus

inferior petrosal sinus

sigmoid sinus

transverse sinus

FIG. 37

The main intracranial venous channels on the base of the skull.

On the posterior surface of the petrous temporal is the **internal acoustic meatus** which transmits the facial and vestibulocochlear nerves and the labyrinthine branch of the basilar artery.

The membranes (meninges) (Fig. 33)

The **dura mater** is the strong outer membrane covering the brain, and the pia and arachnoid membranes form the inner coverings. The **pia** covers the surface of the brain, enters the sulci (grooves) and accompanies the blood vessels into the brain substance. The **arachnoid** is a thin membrane in close contact with the dura and is separated from the pia by the **subarachnoid space** which contains the cerebrospinal fluid. The inner surface of the dura is covered by a mesothelium and the outer surface forms the endocranial lining of the bones of the cranial cavity. The dura is continuous with the pericranium at all sutures and foramina but also fuses with the outer fibrous covering of vessels and nerves as they enter or leave the skull. In the case of the optic nerve the dural sleeve and subarachnoid space are continued as far as the back of the eyeball. An increase of intracranial pressure may be transmitted down this sleeve, block the venous return of the eyeball and give rise to **papilloedema** (swelling and protrusion of the optic disc where the optic nerve leaves the eyeball).

The lower part of the dura has a rich nerve and blood supply. Most of the nerves are branches of the trigeminal. The middle meningeal artery enters the skull through the foramen spinosum. Other vessels to the meninges come from branches of the anterior ethmoidal artery as it lies on the cribriform plate and also from the principal branches of the internal carotid and vertebral arteries. The meningeal vessels also supply the bones of the skull. The veins drain into the large venous sinuses.

The **dural venous sinuses** are lined by endothelium. They communicate with each other, drain the blood from the brain and are tributaries of the internal jugular veins (Fig. 37). Small emissary veins pass from the venous sinuses to the scalp and to the pterygoid, diploic, spinal and other venous plexuses. The blood may flow in either direction in the emissary veins and they are important in relation to infection almost anywhere in the head (Fig. 33).

CHAPTER 10

THE ORBIT AND EYEBALL

INTRODUCTION

THE walls of the orbit consist principally of the following bones; the roof—the frontal and the lesser wing of the sphenoid; the floor—the maxilla ; medially—mainly the orbital plate of the ethmoid ; laterally—the zygomatic bone and the greater wing of the sphenoid. The optic canal is posterior and medial. The superior orbital fissure is between the roof and the lateral wall, and the inferior orbital fissure is between the floor and the lateral wall. Identify on the skull and on the living person the notch in the supra-orbital ridge. On the living person the pulley (the **trochlea**) for the superior oblique muscle is palpable in the upper medial corner of the front of the orbit.

The eyeball is lined by the retina which responds to light stimuli and transmits nerve impulses to the optic nerve. Light enters the eyeball through the transparent cornea and is focused by the cornea and lens on to the retina. The white of the eyeball consists of supporting fibrous tissue and gives attachment to the six extra-ocular muscles which move the eyeball so that the cornea faces in the desired direction. The space between the eyeball and the bony orbital walls is filled with fat.

DISSECTION

(The dissection of the orbit and its contents may be omitted.)

Incise the dura along the edge of the lesser wing of the sphenoid as far as the anterior clinoid process. Strip the fibrous tissue from both surfaces of the process and then carefully remove the process and the adjacent part of the lesser wing. Trace the cranial nerves (3rd, 4th, 6th and ophthalmic 5th) forwards into the orbit (Fig. 36A). The orbital plate of the frontal bone is nibbled away as the dissection proceeds. Lying most posteriorly and immediately under the periosteum is the **trochlear nerve**. It crosses above the muscles of the eyeball, runs medially and enters the upper border of the **superior oblique muscle**. The **frontal nerve** can be traced forwards

100

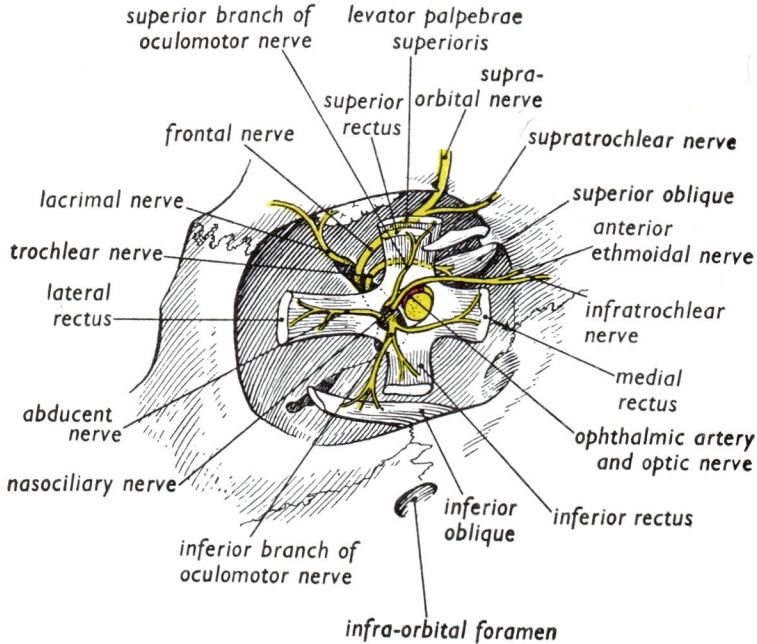

superior branch of oculomotor nerve

levator palpebrae superioris

supra-orbital nerve

superior orbital

frontal nerve

superior rectus

supratrochlear nerve

lacrimal nerve

superior oblique

anterior ethmoidal nerve

trochlear nerve

lateral rectus

infratrochlear nerve

medial rectus

abducent nerve

ophthalmic artery and optic nerve

nasociliary nerve

inferior oblique

inferior rectus

inferior branch of oculomotor nerve

infra-orbital foramen

FIG. 38

Drawing of a dissection of the back of the right orbit after removal of the eyeball. (Compare with Fig. 36A.)

on the surface of the levator palpebrae superioris, and divides into a medial, supratrochlear, and a lateral, supra-orbital branch. Their terminations in the scalp have already been found. The **lacrimal nerve** lies against the periosteum on the lateral side of the orbit and is thin and difficult to dissect. It goes to the lacrimal gland and outer part of the upper eyelid.

The space bounded by the lesser and greater wings of the sphenoid is the **superior orbital fissure** (Fig. 38). Passing through its lateral part, outside the tendinous origin of the orbital muscles, are from lateral to medial the lacrimal, frontal, and trochlear nerves and the superior ophthalmic veins. The veins pass backwards and join the cavernous sinus.

Remove the fat of the orbit piecemeal and define the **levator palpebrae superioris** and, beneath it, the **superior rectus** muscle.

101

Medially the superior oblique can be found and inferior to this muscle is the **medial rectus**. On the lateral side lies the **lateral rectus.** Cut across the levator palpebrae superioris and the superior rectus behind the insertion of the latter into the eyeball. Trace their nerve supply, the superior branch of the oculomotor nerve, into these two muscles. Now find the **optic nerve** emerging from the back of the eyeball, and the **nasociliary** branch of the ophthalmic nerve and the **ophthalmic artery** passing forwards above the optic nerve from the lateral to the medial side. Short branches leave the nasociliary nerve to enter the **ciliary ganglion,** which is small and lies on the lateral side of the optic nerve (Fig. 41).

Pull the levator palpebrae superioris and the superior rectus backwards as far as possible (cutting their nerves if necessary) and cut through the tendinous ring of origin lateral to the levator muscle. The **tendinous ring** passes round the optic canal and the medial part of the superior orbital fissure, and the four recti muscles and the levator muscle all arise from it. The lateral rectus arises by two heads and passing between them is the **abducent nerve** which enters the medial surface of this muscle. The nasociliary nerve can also be seen passing through the ring and inferior to the nasociliary nerve is the inferior branch of the oculomotor nerve. The nasociliary nerve passes forwards and medially over the optic nerve and divides into: (1) the **infratrochlear branch** supplying the skin of the root of the nose and the medial part of the upper eyelid, and (2) the **anterior ethmoidal nerve** running through the anterior ethmoidal foramen into the anterior cranial fossa. The latter nerve passes forwards on the cribriform plate (outside the dura), enters the upper part of the nasal cavity and ends in the skin on the side of the nose. It supplies the upper part of the nasal mucous membrane and the skin of the side and tip of the nose. The **anterior ethmoidal artery** (a branch of the ophthalmic) accompanies the anterior ethmoidal nerve. Cut the optic nerve and turn the eyeball forwards, thus exposing the **inferior rectus** and the **inferior oblique** muscles, with their nerve supply from the inferior branch of the oculomotor. This nerve also gives branches to the medial rectus. Unlike the other muscles, the inferior oblique runs transversely near the front of the orbit.

102

Dissection of the eye of the ox

Open up the interior of two ox eyes, cutting one round the equator in a coronal plane and the other in a vertical sagittal plane. Apart from the size of the ox's eye, the main difference from that of a man is the presence of the highly iridescent **tapetum,** lying between the retina and choroid.

Identify the **lens** and **ciliary body,** the **iris** with the **anterior** and **posterior aqueous chambers,** the **vitreous chamber,** the **cornea** and the **sclera** (Fig. 41). Examine carefully the posterior half of the eye sectioned coronally. Pull the vitreous body to one side and note whether the retina is attached to it. If it is, place the eye in water and float off the vitreous. Identify the blood vessels radiating from the **optic disc.** The outermost coat, the sclera (white of the eye), is made of strong fibrous tissue. Deep to this is the vascular and pigmented **choroid,** then the tapetum (iridescent layer), then the retina. The retina is almost transparent and is the light sensitive layer of the eye. The other coats are protective and nutritive in function. Examine in the other part of the eye the lens capsule which is attached to the ciliary body, and then remove the lens and expose the posterior surface of the iris. Note its highly pigmented nature and the shape of the **pupil** which is very different from the shape in man. The iris divides the space containing the aqueous humor into anterior and posterior chambers. Examine the sagittal section of the eye and identify the various structures and the sub-divisions of the cavity into the chambers.

STRUCTURAL DETAILS

The walls of the orbit

The orbit is roughly a four-sided pyramid with an apex at the back and a base formed by the orbital margin. The roof separates the orbit from the anterior cranial fossa and consists of the orbital plate of the frontal and the lesser wing of the sphenoid bones. The lateral wall separates the orbit from the infratemporal fossa laterally and the middle cranial fossa behind. It consists of the zygoma in front and the greater wing of the sphenoid behind. The floor consists of the zygoma and the orbital plate of the maxilla beneath which is the maxillary sinus. The medial wall is formed from before backwards by the frontal process of the maxilla, the lacrimal bone, the orbital plate of the ethmoid and the body of the sphenoid. It separates the orbit from the nasal cavity with the ethmoidal air sinuses intervening.

The **optic canal** transmits the optic nerve and ophthalmic artery and opens into the apex of the orbit. The **superior orbital fissure** passes laterally and upwards from the apex, between the roof and the lateral wall. It transmits the 3rd, 4th, and 6th cranial nerves and the branches of the ophthalmic division of the 5th nerve, and the ophthalmic veins. A tendinous ring surrounds the optic canal and the medial part of the superior orbital fissure.

The **inferior orbital fissure** extends laterally and downwards from the apex, between the floor and the lateral wall. It is filled with fibromuscular tissue in the living and is pierced by communicating veins between the orbit and the pterygoid plexus, and

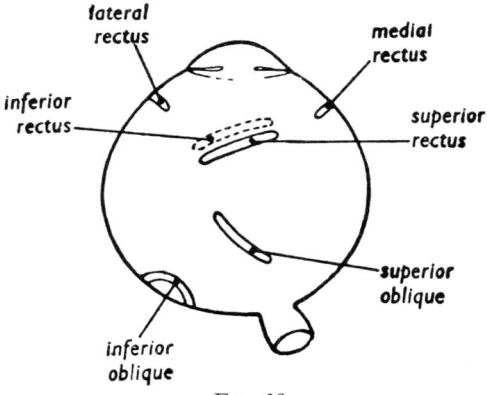

FIG. 39
The left eyeball viewed from above indicating the attachments of the bulbar muscles.

by the zygomatic nerves. The maxillary nerve and artery enter the orbit through this fissure and become the infra-orbital nerve and artery. In the skeleton, the orbit can be seen to communicate through the inferior orbital fissure medially with the pterygopalatine fossa and laterally with the infratemporal fossa.

The bulbar muscles

The bulbar (extra-ocular) muscles are responsible for movements of the eyeball and elevation of the upper lid. The **levator**

palpebrae superioris arises from the tendinous ring and adjacent bone and is attached to the fibrous tarsal plate of the upper lid. The four **recti** muscles arise from the tendinous ring and are inserted on to the eyeball in a circle in front of the equator (Fig. 39). The **superior oblique** arises from bone above and medial to the optic canal. The muscle runs forwards and the tendon passes through a fibrocartilaginous pulley, the **trochlea,** turns laterally and backwards inferior to the superior rectus and is attached to the eyeball behind the equator just lateral to the superior rectus. The **inferior oblique** runs laterally below the inferior rectus from its origin on the antero-medial part of the floor of the orbit to its insertion behind the equator lateral to the optic nerve.

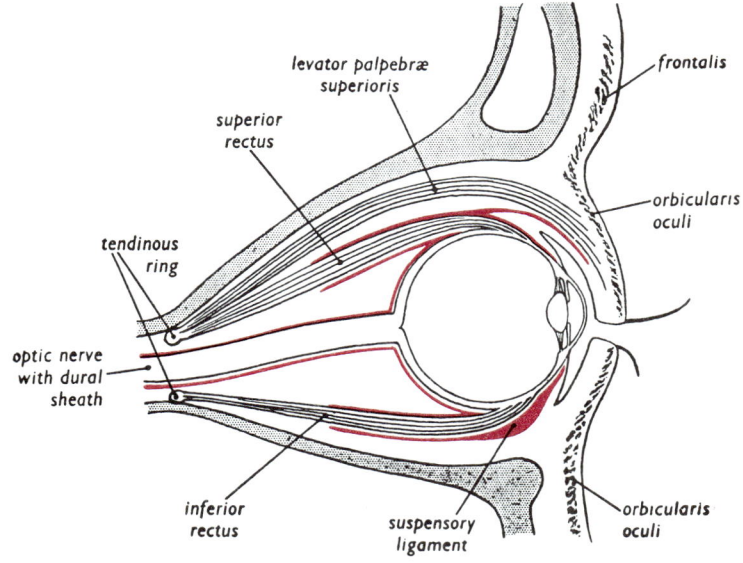

FIG. 40

Diagram of the relations of the orbital fascia outlined in red.

The orbital fascia

The eyeball behind the cornea is covered by a layer of fibrous tissue called the orbital fascia (Fig. 40). The muscles attached to the sclera pass through this fascia, which is prolonged backwards as a sleeve around each muscle. The fascia is specially thickened under the eyeball, where it forms a suspensory ligament attached to the bony walls medially and laterally.

The arteries and veins of the orbit

The orbit is supplied by the **ophthalmic artery,** a branch of the internal carotid. It has named branches accompanying the branches of the ophthalmic nerve and also muscular branches. Another important branch is the **central artery of the retina,** which runs in the optic nerve and pierces the optic disc to supply the retina. It is accompanied by a vein which may be compressed by increased cerebrospinal fluid pressure. Other arteries pierce the sclera in a circle round the optic nerve and supply the choroid, ciliary body and iris.

The ophthalmic veins drain the eyeball and the other contents of the orbit, pass through the superior orbital fissure and enter the cavernous sinus. Some veins pass through the inferior orbital fissure and enter the pterygoid plexus. In front, the ophthalmic veins frequently anastomose with the beginning of the facial vein. Thus infection may pass from the face into the orbit and so to the cavernous sinus.

The nerves in the orbit (Figs. 38, 41 and 57)

The **optic nerve** passes through the optic canal with the ophthalmic artery. It is surrounded by the meningeal sheaths and a prolongation of the subarachnoid space. The central artery of the retina enters a fissure on its inferior side. The optic nerve enters the posterior surface of the eyeball slightly medial to its centre. The site of entry, the **optic disc,** may be seen in the living eye with an ophthalmoscope. The **oculomotor nerve** divides into two branches in the cavernous sinus and passes through the tendinous ring into the orbit. The superior branch supplies the superior rectus and pierces it to supply the levator palpebrae superioris. The inferior branch supplies the medial rectus, the inferior rectus and the inferior oblique. Parasympathetic fibres to the ciliary ganglion travel with the branch to the inferior oblique. The **trochlear nerve** passes through the superior orbital fissure, crosses over the posterior attachment of the levator palpebrae superioris, and enters the upper edge of the superior oblique muscle. The **abducent nerve** enters the orbit through the superior orbital fissure and within the tendinous ring. It supplies the lateral rectus.

The **ophthalmic division of the trigeminal nerve** is responsible for the sensory innervation of the orbital structures and the tissues surrounding the orbit, especially above. The **frontal** branch passes to the forehead and the conjunctiva; the **lacrimal** branch supplies the lateral part of the upper eyelid, and its branches to the lacrimal gland transmit secretomotor fibres from the pterygopalatine ganglion (Fig. 57). The **nasociliary** branch carries sensory impulses from the eyeball and transmits sympathetic fibres to the dilator pupillae from the carotid plexus. Other branches of the nasociliary nerve are the **infratrochlear** (sensory from the medial part of the upper eyelid and the upper part of the external nose), the **anterior ethmoidal** (sensory from the mucous membrane of the upper part of the nasal cavity and from the skin of the lower part and tip of the external nose), and a branch to the ciliary ganglion.

The ciliary ganglion

The ciliary ganglion contains the synapses of the parasympathetic nerve supply to the eyeball (Fig. 41). The preganglionic branches come from the oculomotor nerve and the postganglionic branches pass to the eyeball to supply the ciliary and sphincter pupillae muscles. Stimulation of these parasympathetic fibres causes (*a*) the sphincter pupillae to contract, so that the aperture of the iris becomes smaller, and (*b*) contraction of the ciliary muscles so that the capsule of the lens relaxes and the lens becomes more biconvex (accommodation). Other nerves, both sympathetic (from the carotid plexus) and sensory (from the nasociliary nerve) pass through the ganglion but do not synapse there. The sympathetic fibres supply the vessels in the eyeball.

The eyeball (Fig. 41)

The eyeball is roughly spherical in shape and consists of the dense white supporting tissue, the **sclera** which is specially modified in front to form the transparent window, the **cornea**. The vascular and pigmented **choroid** lines the inner surface of the sclera and lining the choroid is the light sensitive **retina**. Round the region of the sclerocorneal junction on the inside is the **ciliary body** which contains smooth muscle. The **lens** lies behind the cornea and is attached to the ciliary body by the **suspensory ligament** (zonular fibres). A circular projection from the ciliary body, the **iris**, passes

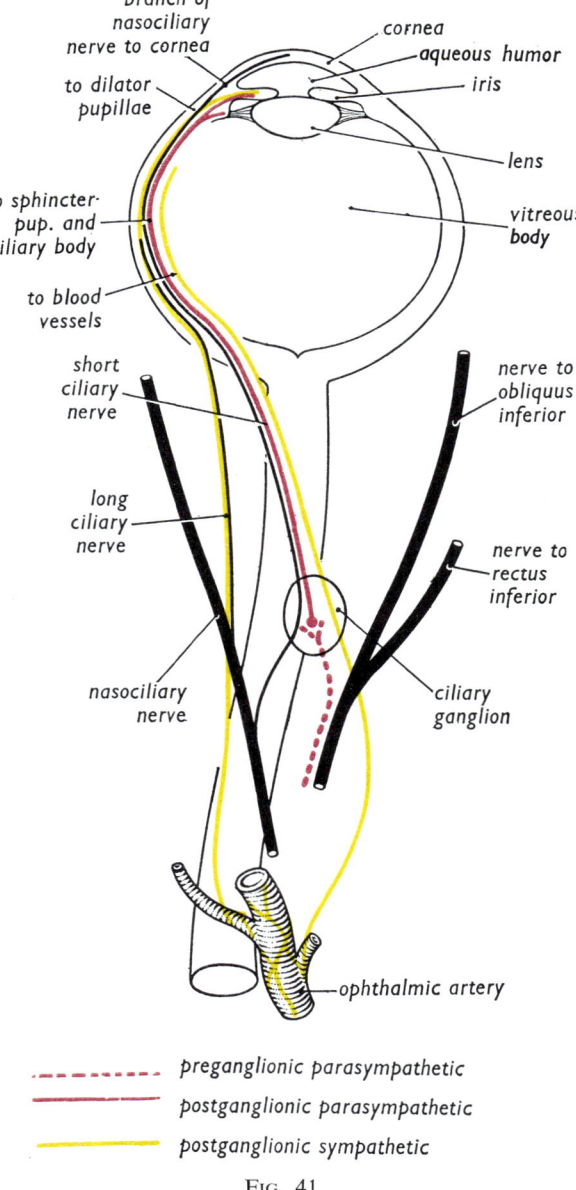

branch of
nasociliary
nerve to cornea

cornea

aqueous humor

to dilator
pupillae

iris

lens

to sphincter·
pup. and
ciliary body

vitreous
body

to blood
vessels

short
ciliary
nerve

nerve to
obliquus
inferior

long
ciliary
nerve

nerve to
rectus
inferior

nasociliary
nerve

ciliary
ganglion

ophthalmic artery

preganglionic parasympathetic

postganglionic parasympathetic

postganglionic sympathetic

FIG. 41
Diagram of the position and connexions of the
right ciliary ganglion.

108

in front of the lens and partially divides the space between the cornea and the lens into the **anterior** and **posterior chambers.** The iris has a circular central aperture, the **pupil.** The iris contains smooth muscle, blood vessels and pigment. The anterior and posterior chambers of the eye contain a watery fluid, the **aqueous humor** which is constantly formed from vessels of the ciliary body and re-absorbed in other vessels. The space behind the lens is filled by a viscid fluid called the **vitreous body** which cannot be replaced if lost.

The light sensitive cells of the retina are the **rods** and **cones.** Impulses originating in these cells are transmitted via bipolar cells to ganglion cells whose axons lie on the inner surface of the retina adjacent to the vitreous body. These fibres converge on the optic disc where there are no light sensitive cells so that a " blind spot " is formed. Lying lateral to the optic disc, where the visual axis of the eye projects on to the retina, is the **macula,** the area specially used in daylight vision.

Sympathetic nerves supply the dilator muscle of the iris. Parasympathetic nerves supply the circular muscle of the iris and the muscle of the ciliary body and when these muscles contract the pupil becomes smaller and the suspensory ligament of the lens is relaxed.

The conjunctiva

This is a thin movable mucous membrane lining the inner surface of the eyelids and continued on to the sclera. At the sclerocorneal junction it becomes very thin and is firmly attached to the cornea. The sensory fibres from the cornea and the region round the sclerocorneal junction run in the nasociliary nerve. The remainder of the scleral and lid conjunctiva is supplied by the supratrochlear, supra-orbital and lacrimal nerves above, the infratrochlear nerve medially and the infra-orbital nerve below.

The lacrimal apparatus

Tears are produced by the lacrimal gland, situated in the superolateral side of the orbit and in the lateral part of the upper eyelid. The fluid enters the conjunctival sac through numerous small ducts. Most of the fluid evaporates but some enters the lacrimal canals through the lacrimal puncta, passes to the lacrimal sac and thence via the nasolacrimal duct to the inferior meatus of the nasal cavity.

The tarsal glands

The eyelids contain many specialised glands as well as the glands in the skin and the conjunctiva. The large **tarsal (Meibomian) glands** lies deep to the conjunctiva, open on the edge of the lid behind the lashes and can be seen when the lid is everted. They are sebaceous glands but are not connected with hair follicles. The sebaceous material lines the edges of the eyelids and seals the palpebral fissure when the lids are closed. It also forms a thin film over the exposed surface of the open eye.

FUNCTIONAL ASPECTS

Movements of the eyeball

By means of the co-ordinated action of the muscles the visual axes are correctly orientated as the eyes are moved to fix on different objects. Movements around horizontal and vertical axes (up and down and side to side) are brought about mainly by the

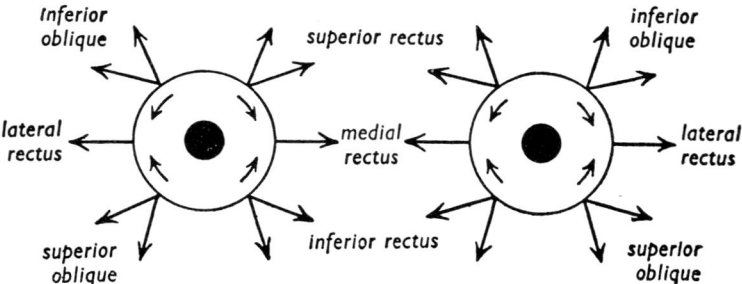

FIG. 42
Diagram indicating the actions of the eye muscles. The small curved arrows indicate rotation of the eyeball.

recti and oblique muscles. It must be emphasised, however, that movements are brought about by the co-ordinated activity of all the muscles working together.

Figure 42 is a diagrammatic representation of the movements produced by the extra-ocular muscles, *e.g.* the superior oblique assists in directing the pupil downwards and outwards and the inferior rectus directs the pupil down and in. They also rotate the eyeball about a longitudinal axis in the directions indicated by the small arrows.

CHAPTER 11

THE EAR

INTRODUCTION

WITHIN the ear are two mechanisms that partly overlap morphologically and functionally. The **vestibular** mechanism reacts to changes in the position of the head in space and the **cochlear** mechanism reacts to sound vibrations. These two types of receptors send impulses to the brain along the two parts of the vestibulocochlear nerve.

The end-organs for hearing and balance are found in the **membranous labyrinth** within the petrous part of the temporal bone. The membranous labyrinth develops from the **otic vesicle,** an invagination of the surface ectoderm at the side of the hind brain in the early embryo. The otic vesicle contains **endolymph,** a clear watery fluid filling the entire closed system. The embryonic otic vesicle develops into the cochlear duct, a spiral tube for sound detection, and the vestibular apparatus, a series of tubes and sacs providing information used for balancing. The end-organs of hearing and balance are bathed in endolymph. Movement of endolymph affects these organs. The membranous labyrinth lies in a space within the petrous temporal bone which conforms to its general shape and is called the osseous labyrinth. The space between the membranous and osseous labyrinths is filled by fluid called the **perilymph** which communicates with the cerebrospinal fluid in the subarachnoid space.

Sound waves entering the external ear eventually stimulate the end organs in the cochlear duct. It is also possible for the temporal bone to transmit the vibration of a tuning fork to the end organs. Normally, however, sound waves pass down the **external acoustic meatus** (Fig. 43) which together with the **auricle** form the **external ear.** This is separated by the **tympanic membrane** (ear drum) from the **middle ear** (tympanic cavity). This is a narrow cavity in the temporal bone, continuous with the upper pharynx through the **auditory (Eustachian) tube.** The middle ear and the tube are developed from the 1st pharyngeal pouch in the embryo. Lying in

111

the middle ear are three small **ossicles.** They are derived from the upper ends of the 1st and 2nd pharyngeal arch cartilages. The 1st cartilage gives rise to the **malleus** and the **incus,** and the 2nd to the **stapes.** These ossicles are outside the mucous membrane lining the cavity of the middle ear. The malleus is attached to the ear drum and articulates with the incus, which in turn articulates with the

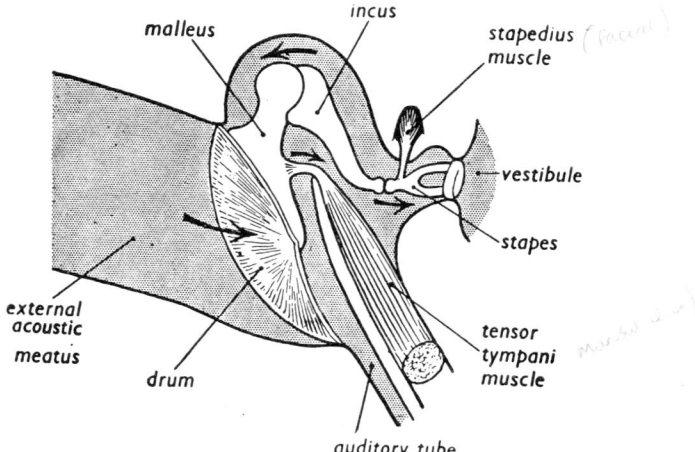

FIG. 43

Diagram of the relations of the ossicles of the ear. The large arrows indicate the movements of the ossicles in response to pressure on the outside of the drum.

stapes. The footpiece of the stapes fills a small opening on the lateral wall of the bony internal ear. Vibration of the drum is thus transmitted through the ossicles to the perilymph, and so to the end-organ, the spiral organ of the cochlear duct.

On the lateral wall of the skull identify the external acoustic meatus. This is surrounded, except posterosuperiorly, by the **tympanic part** of the temporal bone. The defect in the plate is completed by part of the squamous temporal bone which projects downwards behind the meatus. Inside the skull, the petrous part of the temporal bone forms a wedge between the greater wing of the sphenoid in front, and the basilar part of the occipital and the sphenoid behind and medially. The apex of the petrous temporal is separated from the sphenoid by the irregular foramen lacerum.

112

DISSECTION

(This dissection may be omitted.)

Clear the muscles from the mastoid process and with a chisel open up the air cells. Remove as many of these as possible and define the larger space of the **mastoid (tympanic) antrum** (Fig. 44). Remove the posterior wall of the external acoustic meatus as

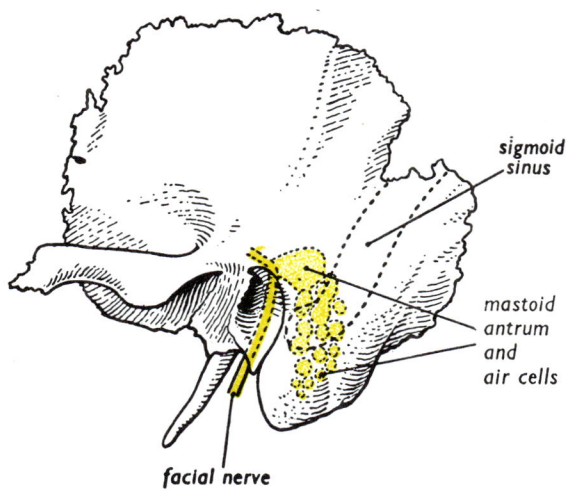

FIG. 44

The left temporal bone. The position of the facial nerve, the sigmoid sinus, the mastoid antrum and the mastoid air cells is indicated.

far as the attachment of the tympanic membrane. Pulling the auricle upwards and backwards brings the cartilaginous part in line with the osseous part since the meatus is curved upwards and backwards. Attached to the inside of the membrane is the handle of the malleus which passes downwards and backwards.

Remove with a saw the part of the squamous portion of the temporal bone which forms the lateral wall of the middle fossa of the skull. Identify the **arcuate eminence** on the petrous temporal indicating the position of the anterior semicircular canal. The bone anterolateral to this eminence is the roof of the middle ear and antrum (the **tegmen tympani).** With a chisel carefully cut off the

113

upper part of the eminence, thus exposing the bony canal and more laterally, remove the roof thus opening the middle ear and the tympanic antrum. Note the close proximity of the **sigmoid sinus** to the antrum and the thinness of the bone separating the antrum from the middle and posterior cranial fossae.

Now look at the middle ear from above and from the side. Above the drum is a small epitympanic recess and in it is the head of the malleus articulating with the body of the incus. The stapes can be seen lying between the lower end of the incus and the medial wall of the middle ear. Passing forwards between the handle of the malleus laterally and the incus medially is the chorda tympani nerve. Carefully press on the ear drum with forceps or seeker and note the movements of the ossicles.

Remove all the bone surrounding the anterior semicircular canal and note that the medial end of the anterior canal joins with the upper end of the posterior canal. These canals lie in vertical planes at right angles to each other.

Chisel away the roof of the **internal acoustic meatus** and trace the **facial, cochlear** and **vestibular nerves** through it. The facial (7th) nerve is most superior and can be traced laterally across the inner ear to a point on the medial wall of the middle ear, where it turns sharply backwards along the junction of the roof and medial wall (Fig. 45). At the point where the nerve turns backwards is the **geniculate ganglion** of the facial nerve. Running forwards from the ganglion is the **greater petrosal nerve** (Fig. 31). Cut the facial nerve in the internal acoustic meatus and turn it laterally, thus exposing the cochlear and vestibular divisions of the 8th cranial nerve. Carefully chip away the bone between the internal acoustic meatus and the greater petrosal nerve. The coils of the cochlea will be exposed. Note that the cavity of the cochlea is partially subdivided by a **spiral lamina** winding round the **modiolus** (central pillar). Gradually remove the bone below the ganglion of the facial nerve and notice that the basal coil of the cochlea bulges into the middle ear as the **promontory.** Before the level of the promontory is reached, the stapes will have come clearly into view with its base (the footpiece) lying in the **fenestra vestibuli** (foramen ovale). The base of the stapes closes the foramen. The fenestra vestibuli lies immediately below the canal for the facial nerve. Running to the

114

stapes from the posterior wall of the middle ear is the tendon of the **stapedius** muscle. Chip away more of the bone in front of the middle ear and expose the **tensor tympani** muscle, whose tendon is attached laterally to the handle of the malleus (Figs. 43 and 45). The stapedius and the tensor tympani probably diminish the sensitivity of the ossicles to low frequency sounds.

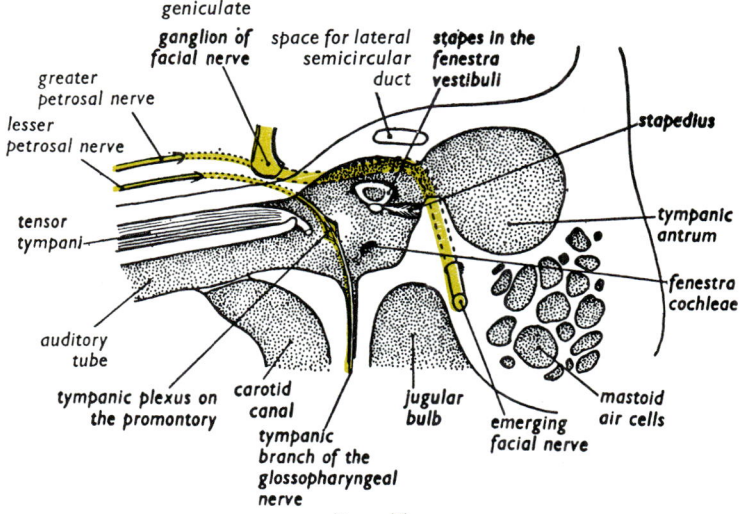

FIG. 45

Diagram of the structures seen in the left middle ear. The lateral wall has been removed.

Pass a wire forwards and medially from the middle ear down the **auditory tube** as it runs parallel to and below the canal for the tensor tympani. Expose the internal carotid artery lying in the bone medial to the tensor tympani and trace the artery across the foramen lacerum to the cavernous sinus.

Trace the cochlear nerve into the modiolus (central pillar) of the bony cochlea and the vestibular nerve into the lateral region of the vestibule which is the space in the bone between the semicircular canals behind and the bony cochlea in front. One end of a bony semicircular canal is enlarged. The cavity of the bony semicircular canals is continuous with that of the vestibule. Many of the vestibular nerve endings are in the **ampullae** of the membranous **semicircular ducts** which lie in the bony canals.

115

STRUCTURAL DETAILS

The temporal bone

The temporal bone ossifies from a number of different centres. Some of these are cartilage bones which form in the wall of the auditory capsule. They fuse to make the central **(petromastoid)** part of the temporal and enclose the internal ear. The **tympanic** part appears as a separate ring in membrane and forms part of the external acoustic meatus. The **squamous** part of the temporal is a membrane bone and forms part of the wall of the cranium and part of the zygomatic arch. The **styloid process** is a portion of the hyoid (2nd pharyngeal) arch and is a cartilage bone. The whole temporal bone can be described in terms of the above four main developmental parts.

(1) The **petromastoid part** of the temporal, containing the internal ear, lies in the base of the skull between the greater wing of the sphenoid in front and the occipital bone behind. On the intracranial surface it forms part of the middle and posterior cranial fossae. The **mastoid process** is readily palpable behind the auricle, and the sternocleidomastoid and other muscles are attached to it. The upper part of the mastoid process contains an air space, the **mastoid antrum**. Continuous with this in the remainder of the mastoid process are the **mastoid air cells**. These air cells do not extend as far as its tip.

At birth the mastoid process is very small. The stylomastoid foramen is therefore almost on the lateral surface of the skull and the facial nerve is relatively superficial as it emerges from its foramen. The **mastoid emissary vein** passes through the bone and connects the transverse sinus with the posterior auricular vein.

(2) The **tympanic part** of the temporal bone forms the main bony part of the external acoustic meatus and also contributes to the bone behind the articular area. The deficiency in the meatus above and behind is filled by the squamous part.

(3) The **squamous part** is a plate of bone at the side of the skull posterior to the greater wing of the sphenoid. The **zygomatic process** of the temporal bone, which forms more than half the zygomatic arch, projects forwards from the lower part of the outer surface of the squamous temporal. The squamous temporal forms

116

the **articular tubercle** and the **mandibular fossa** for the condylar process of the mandible. Posterior to the articular area is the **squamotympanic fissure** which is subdivided by a thin wedge of petrous bone. The chorda tympani emerges from the **petrotympanic fissure.**

(4) The **styloid process** is long and slender and projects downwards from the inferior surface of the bone immediately behind the lower margin of the tympanic plate. In the intact skull it is usually 2-3 cm long.

The external and internal acoustic meati are easily identified but other important foramina exist in the temporal bone especially on its inferior surface. Between the styloid and mastoid processes is the **stylomastoid foramen** out of which comes the facial nerve. On the inferior surface of the petrous temporal is the **carotid canal** separated by a layer of bone from the posteriorly placed **jugular fossa.** At the apex of the petrous temporal the openings of the canals for the tensor tympani (above) and the auditory tube (below) can be seen. On the upper anterior surface of the petrous temporal are openings for the greater and lesser petrosal nerves. The lesser is more inferior and lateral.

The external ear

The external ear includes the auricle and the meatus. The auricle consists of elastic cartilage, has an irregular shape and directs the sound waves into the meatus. The most dependent part is the **lobule** and in front of the meatus is the **tragus.** Attached to the auricle are some vestigial muscles.

The **external acoustic meatus** is a partly cartilaginous and partly bony tube leading to the ear drum. It is developed from the 1st pharyngeal cleft and is lined by skin containing ceruminous glands. Excessive secretion of these may block the meatus and lead to deafness. The direction of the meatus changes, being slightly backwards in the outer cartilaginous part but slightly forwards in the inner bony part. The tympanic part of the temporal bone forms most of the bony meatus. The **tympanic membrane (ear drum)** is elliptical in shape and slopes downwards and inwards in such a way that the postero superior quadrant is most lateral and the antero-inferior quadrant is most medial. Developmentally the drum represents the membrane between the 1st pharyngeal cleft (formed of ectoderm) and the 1st pharyngeal pouch (formed of endoderm). Between the two epithelial layers is a packing of fibrous tissue but this is almost absent from the upper part of the membrane, known as the **flaccid part.** The handle of the malleus passes downwards and backwards and is firmly attached to the drum about its middle. The lower end of the handle produces a dimple in the membrane and from this point, when the drum is inspected with an auroscope, a cone of light is seen running downwards and forwards. Movement of the drum can be seen during swallowing, if the auditory tube is patent. Changes in the colour of the drum may serve to indicate disease in the middle ear.

The middle ear

The middle ear is a small space in the temporal bone about 15 mm in its anteroposterior and vertical diameters and 2 mm at its narrowest from side to side (Fig. 45). It lies in the petrous part of the temporal bone but its lateral wall is formed by the squamous temporal and the tympanic part. This indicates that the upper end of the first pharyngeal pouch is enclosed except laterally by the petrous temporal (Fig. 46). The cavity of the middle ear is lined by ciliated mucous columnar epithelium, continuous with that of the pharynx through the auditory tube. The cavity extends backwards through a small opening into the tympanic (mastoid) antrum (Fig. 45). Infection in the pharynx may spread along the auditory tube into the middle ear and mastoid antrum.

The lateral wall is largely filled by the tympanic membrane. Immediately above and behind the membrane is a recess containing the head of the malleus and the body of the incus. This recess leads into the antrum through an opening, the **aditus.** The handle of the malleus, attached to the membrane, and the long process of the incus, behind the handle, extend down into the middle ear. Passing forwards between them is the chorda tympani nerve.

The medial wall has at about its middle a bulge, the **promontory,** which is about 2 mm from the centre of the drum. This bulge is caused by the basal coil of the cochlea. Above and behind the promontory is the **fenestra vestibuli** (foramen ovale) and below and behind is the **fenestra cochleae** (foramen rotundum) (Fig. 45). Both these windows open into the vestibule, the central part of the osseous labyrinth. The fenestra vestibuli is closed by the base (foot-piece) of the stapes and the fenestra cochleae by a fibrous disc. These two openings transmit vibrations to or from the perilymph in the vestibule. Between the roof and the medial wall is the ridge formed by the canal for the facial nerve. Posteriorly the canal turns downwards medial to the aditus and runs between the medial and

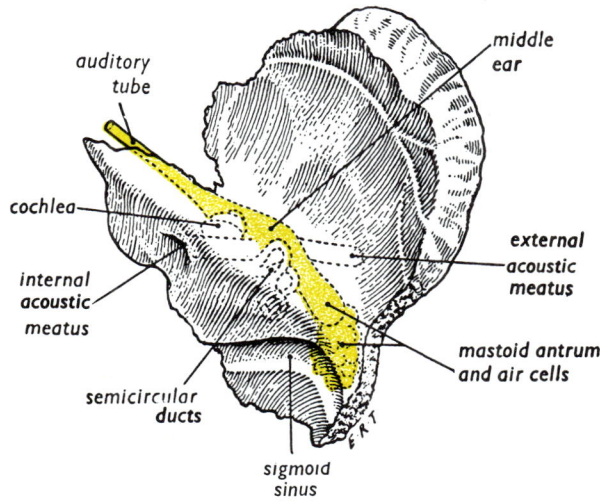

FIG. 46

The right temporal bone viewed from above. The relative positions of some of the structures are projected on to the surface.

posterior walls to the floor. The nerve passes out of the temporal bone at the stylomastoid foramen (Fig. 31). Pressure on the facial nerve in its bony canal due to an inflammatory process may cause paralysis of the muscles of facial expression. During operations on the middle ear the nerve in its canal is at risk.

The posterior wall is taken up mostly by the aditus to the antrum, medial to which is the ridge formed by the facial canal. Below the aditus is the pyramidal eminence, out of which emerges the stapedius muscle (supplied by the facial nerve and attached to the stapes). Lateral to the pyramid is the small posterior foramen from which the chorda tympani emerges.

The openings of two canals, separated by a shelf of bone, are on the anterior wall (Fig. 45). The upper canal contains the tensor tympani muscle (supplied by the mandibular nerve). Its tendon turns at right angles round the posterior end of the shelf and is attached laterally to the handle of the malleus. The lower opening is the lateral end of the auditory tube, the upper bony part of which is about 1·3 cm long. The tube runs downwards, medially and forwards to the lateral wall of the nasopharynx and is cartilaginous in its medial part.

The floor of the middle ear is formed by a thin plate of bone separating it from the superior bulb of the internal jugular vein and in front of this is the carotid canal.

The roof is formed by a thin plate of bone which separates the middle ear from the middle cranial fossa, the meninges and the temporal lobe of the brain.

The mastoid (tympanic) antrum

This space in the mastoid portion of the temporal bone is present at birth, when it is very superficial, but it becomes gradually deeper as the mastoid process increases in size during the first two years of life. In front, it is continuous with the middle ear and on the medial wall are seen the ridges covering the facial canal in front and the lateral semicircular canal behind. Posteromedially a thin plate of bone separates the antrum from the transverse venous sinus beyond which is the cerebellum in the posterior fossa of the skull. Above, the antrum is separated by thin bone from the meninges and the temporal lobe of the cerebrum (Fig. 47).

The antrum is formed by an evagination backwards from the middle ear and is lined by squamous epithelium. After birth air cells develop around it in the enlarging mastoid process.

Infection from the throat may spread to the middle ear and from there into the tympanic antrum and air cells (mastoiditis). This may result in perforation of the drum, or infection may spread among the air cells towards the tip of the mastoid process, or into the posterior cranial fossa and the cerebellum, or into the middle cranial fossa and the temporal lobe of the cerebrum (Fig. 47).

The internal ear (Fig. 48)

The internal ear contains the membranous labyrinth which consists of the duct of the cochlea, the utricle, the saccule, and the

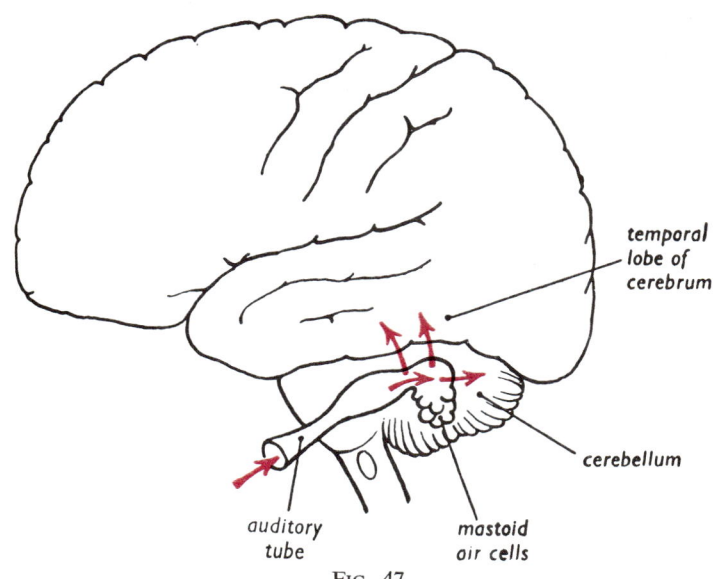

temporal
lobe of
cerebrum

cerebellum

auditory
tube

mastoid
air cells

FIG. 47

Diagram of the possible spread of infection from the pharynx to
the middle ear, mastoid antrum and the brain.

semicircular ducts with their small communicating ducts. It also
contains the end organs associated with the peripheral processes of
the vestibulocochlear nerve.

A. The cochlea

The membranous duct of the cochlea is coiled round the
modiolus. Lying on the basilar membrane (which forms the floor
of the duct of the cochlea) is the **spiral organ.** This contains the
peripheral endings of the cochlear nerve which are indirectly stimu-
lated by movements of the basilar membrane which vibrates due to
waves in the perilymph produced by the incoming sounds. On
both sides of this endolymph duct there is a coiled perilymph
canal. The upper of the perilymph coils is the **scala vestibuli** and
it begins opposite the fenestra vestibuli in which is the base of the
stapes. This coil is larger at the base of the spiral and smaller at
the apex where it is continuous through an opening, called the
helicotrema, with the **scala tympani.** This latter coil increases
towards the base and ends opposite the fenestra cochleae which has

121

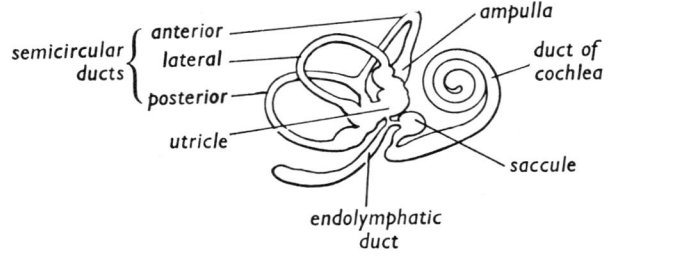

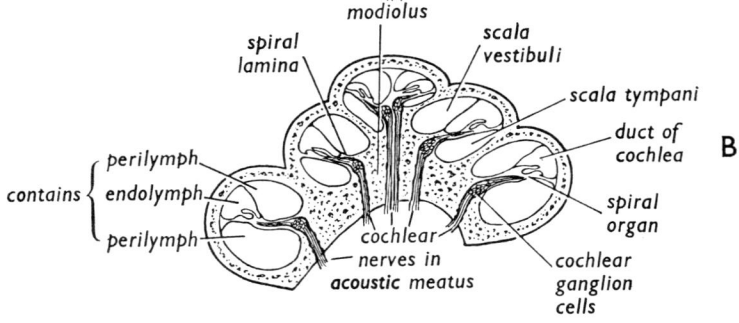

FIG. 48

A. Diagram of the membranous labyrinth. B. Diagram of a section of the bony cochlea, showing some of the relations of the duct of the cochlea and the spiral organ.

a fibrous occluding disc. The large basal turn of the bony cochlea opens into a space in the bone called the **vestibule** in whose lateral wall are the fenestra vestibuli and the fenestra cochleae. In the vestibule are parts of the membranous labyrinth—the **saccule** joined to the membranous cochlea, and the **utricle** out of which pass the three semicircular ducts. The utricle and the saccule are joined together by a small duct.

B. The semicircular ducts

The three ducts lie in the bony semicircular canals of the temporal bone and communicate by five openings with the utricle. The **anterior** and **posterior** ducts lie in vertical planes at right angles to each other. One end of the anterior duct and one end of the posterior duct have a common opening into the utricle. The anterior duct lies in an anterolateral plane and the posterior in a postero-lateral plane. The **lateral** duct is in a horizontal plane. One end of each duct is dilated (the **ampulla**) and situated there is the end-organ of balance (the **ampullary crest**). In the utricle and the saccule similar end-organs (called **maculae**) are present.

The membranous ducts contain endolymph and are bathed in perilymph in the bony canals. In the endolymph are small crystals of calcium carbonate (called **otoliths**) adherent to the hair cells of the maculae. Displacement of fluid and movements of the otoliths due to movement of the head stimulate the hair cells and the ends of the vestibular nerves found around the hair cells. The impulses set up in the nerves pass to the hindbrain.

C. The vestibulocochlear (8th) nerve

Nerve fibres from the end organs of balance and hearing form the vestibulocochlear nerve. The cell bodies of the cochlear fibres are in the spiral ganglion in the modiolus of the cochlea, and those of the vestibular fibres are in a ganglion in the internal acoustic meatus. The vestibulocochlear and facial nerves can be followed from the meatus to the lower border of the pons.

FUNCTIONAL ASPECTS

Deafness may be caused by peripheral or central defects. The former include blockage of the external acoustic meatus by wax or a foreign body, absorption of air from the middle ear due to a blockage of the auditory tube, inflammatory or vascular disease of the middle ear, and diseases of the cochlea. Central defects include diseases of the cochlear nerve or its central pathways in the brain. Diseases of the labyrinth give rise to subjective feelings of giddiness and upset the body balancing mechanisms.

CHAPTER 12

THE INFERIOR SURFACE OF THE BASE OF THE SKULL

BEFORE proceeding further it is necessary to examine the base of the skull, many features of which have already been identified (Figs. 1 and 18). Remove the mandible. In the front is the hard palate. It is bounded by the alveolar processes of the maxillae, is at a lower level than the remainder of the base and is separated from the floor of the anterior cranial fossa by the nasal cavity. The anterior two-thirds of the hard palate are formed by the palatine processes of the maxillae and the remainder by the horizontal plates of the palatine bones. In the midline anteriorly is the **incisive canal** which communicates with the nasal cavity. The **greater palatine foramen,** situated at the posterolateral corner of the palate, transmits the greater palatine nerve and vessels from the pterygopalatine fossa to the palate. Immediately behind it are one or two **lesser palatine foramina,** through which the lesser palatine vessels and nerves pass.

The alveolar processes of the maxillae have sockets for the roots of the eight teeth on each side in the permanent dentition. The teeth are described with those of the lower jaw (page 54).

Above the posterior edge of the palate are the posterior apertures **(choanae)** of the nasal cavity—two rectangular openings by which the nasal cavity communicates with the nasopharynx. Each measures about 3 cm vertically and 1 cm transversely. Lateral to the posterior apertures is the pterygoid fossa, bounded by the medial and lateral pterygoid plates of the sphenoid. Below, the medial pterygoid plate projects downwards as the **hamulus,** round which the tendon of the tensor veli palatini hooks to enter the soft palate. The lateral surface of the lateral pterygoid plate is continuous with the infratemporal surface of the greater wing of the sphenoid. Laterally the maxilla is separated from the lateral pterygoid plate by the pterygomaxillary fissure and posterosuperiorly from the greater wing by the inferior orbital fissure. The **foramen ovale,** for transmission of the mandibular division of the trigeminal

124

nerve, lies behind the root of the lateral pterygoid plate. Postero-lateral to the foramen ovale is the **foramen spinosum,** behind which is the **spine of the sphenoid.** The **foramen lacerum** is situated medial to these two foramina and the apex of the petrous temporal bone. The **mandibular fossa,** for the head of the mandible, is seen lateral to the spine of the sphenoid. The fossa is bounded anteriorly by the **articular tubercle** on which the head of the mandible rides when the mouth opens. The fossa is limited posteriorly by the **squamotympanic fissure.** It is through this fissure that the chorda tympani nerve emerges.

The rough inferior surface of the **petrous temporal** bone is seen between the greater wing of the sphenoid in front and the basilar part of the occipital bone behind. Its medial end, the apex, is separated from the sphenoid bone by the irregular foramen lacerum. The lower opening of the **carotid canal** in the petrous temporal is situated behind the spine of the sphenoid. The canal passes upwards, then medially and forwards and opens on to the upper part of the foramen lacerum. (Confirm this by passing a bent probe through the canal.) The **jugular foramen** lies posterior to the canal. The **styloid process** is lateral to the jugular foramen and anteromedial to the **mastoid process.** The **stylomastoid foramen,** which transmits the facial nerve, is seen between the two processes. The **mastoid notch,** for the attachment of the posterior belly of the digastric muscle, is immediately medial to the mastoid process and a narrow groove for the occipital artery lies medial to the notch.

The **foramen magnum** is wider behind than in front since the **occipital condyles,** which articulate with the atlas, encroach upon it in front. The opening of the **hypoglossal canal** lies lateral to the front part of the condyle. Behind the condyle there is usually a fossa into which opens the **condylar canal.**

The bone in front of the foramen magnum is formed by the basilar part of the occipital and the body of the sphenoid bones which, till the age of twenty-five, are separated by cartilage. Find the centrally placed **pharyngeal tubercle** on the occipital bone.

CHAPTER 13

THE PHARYNX AND RELATED STRUCTURES

INTRODUCTION

THE cavity of the pharynx is subdivided for convenience of description into the **nasopharynx** behind the nasal cavity, the **oropharynx** behind the mouth and the **laryngopharynx** behind the larynx. It is lined by mucous membrane. The nasopharynx is covered by ciliated columnar epithelium, and the rest of the pharynx by stratified squamous epithelium. The oropharynx forms a passage common to both the respiratory and alimentary systems. Later in this section the mechanism for closing off the air passages during the swallowing of food will be discussed.

On the skull identify the hamulus at the lower end of the medial pterygoid plate, the posterior end of the mylohyoid line of the mandible and the pharyngeal tubercle. Note that the opening of the auditory tube is posterolateral to the medial pterygoid plate.

The **thyroid cartilage** can be felt in the midline of the neck. Its anterior border, where the two laminae meet, forms the **laryngeal prominence** ("Adam's apple"). Below the thyroid cartilage the arch of the **cricoid cartilage** can be felt. The posterior part of the cricoid cartilage expands to form a flattened lamina projecting upwards between the divergent laminae of the thyroid cartilage.

DISSECTION

(This dissection may be omitted.)

Cut through the neck at the disc between the 7th cervical and 1st thoracic vertebrae, and remove the head and neck from the trunk. Pull the trachea, oesophagus, large vessels and nerves forwards and find behind them the layer of loose areolar tissue which separates these structures from the prevertebral muscles. This areolar tissue is the prevertebral fascia. It passes upwards to the **retropharyngeal space** and can be easily separated as far as the base of the skull. Divide the base of the skull. First map out the lines of section on a dried skull,

126

then, on the cadaver make a series of saw cuts and join them with the help of a chisel and mallet. Begin just behind the mastoid process, then pass obliquely forwards and medially behind the jugular foramen to the side of the basilar part of the occipital bone, then transversely across this bone between the pharyngeal tubercle and the foramen magnum, and then obliquely backwards and laterally, posterior to the jugular foramen and the mastoid. Cut through any structures connecting the two parts of the skull. Keep the posterior part of the skull and attached vertebral column for future study and proceed with the dissection of the anterior part.

The constrictor muscles of the pharynx can now be identified and their main relations seen (Fig. 49). Lateral to the pharynx, just below the base of the skull, are the sympathetic trunk, the internal carotid artery, the vagus nerve and the internal jugular vein in that order from the medial to the lateral side. The hypoglossal nerve is posterior to the vagus and the spinal accessory nerve is seen passing laterally in front of the internal jugular vein (Fig. 17). Inferiorly, the pharynx narrows as it becomes continuous with the oesophagus.

The three constrictor muscles of the pharynx must now be cleaned and examined. The **superior constrictor** is attached to the hamulus of the medial pterygoid plate, to the mandible and to the pterygomandibular raphe between them, the **middle constrictor** to the hyoid bone, and the **inferior constrictor** to the thyroid and cricoid cartilages. The fibres fan out as they sweep backwards and then medially, and are inserted into a fibrous median raphe which is attached above to the pharyngeal tubercle and below merges with the fibrous covering of the oesophagus. The lowest fibres of the superior constrictor lie internal to the upper fibres of the middle constrictor, and there is a similar arrangement between the middle and inferior muscles. During the dissection the **pharyngeal plexus** formed by branches of the glossopharyngeal, vagus and and cervical sympathetic nerves is seen on the outer surface of the middle constrictor.

Examine the muscular spaces related to the three constrictor muscles. Between the upper border of the superior constrictor and the base of the skull the **auditory tube** runs medially and somewhat downwards and forwards, with the **tensor veli palatini**

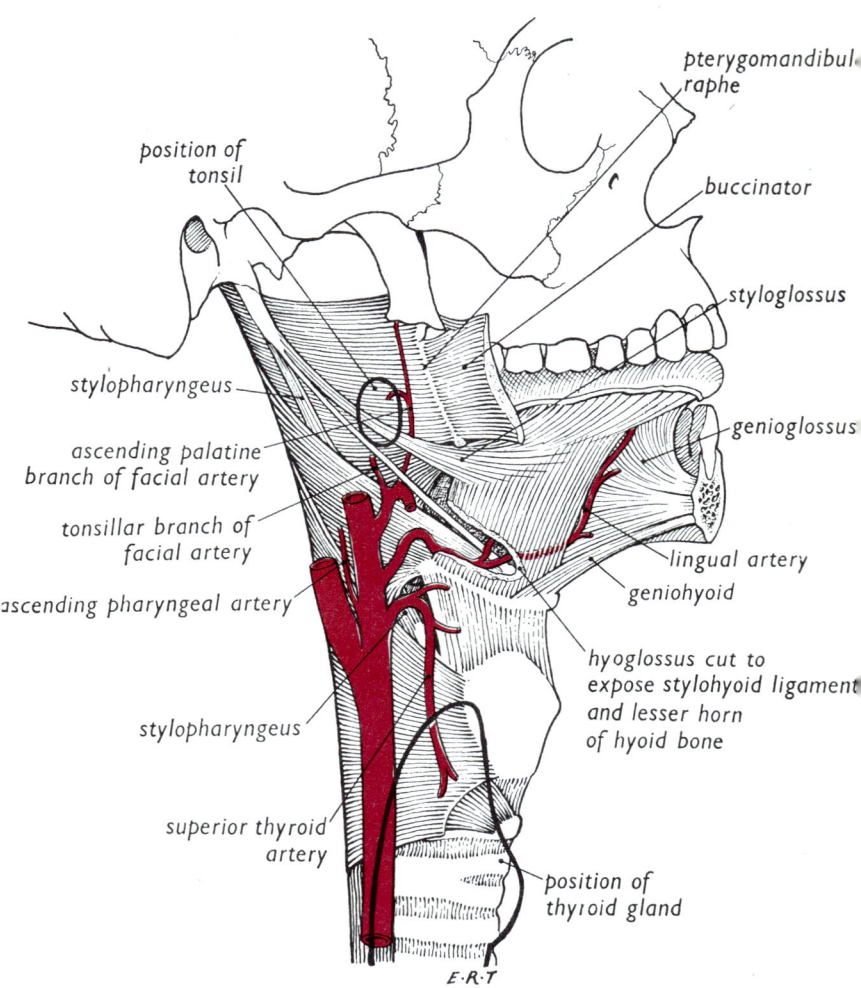

FIG. 49

The attachments of the muscles of the pharynx are indicated along with their principal arterial relations.

muscle laterally and **levator veli palatini** medially. Between the superior and middle constrictors find the **stylopharyngeus** muscle with the glossopharyngeal nerve along its posterior border. Between the anterior parts of the middle and inferior constrictors find the **internal laryngeal nerve** (from the superior laryngeal branch of the vagus) as it pierces the thyrohyoid membrane accompanied by the superior laryngeal branch of the superior thyroid artery. The **external laryngeal branch** of the superior laryngeal nerve, along with the superior thyroid artery, can be traced across the inferior constrictor to the cricothyroid muscle. The **recurrent laryngeal branch** of the vagus and the inferior laryngeal artery (a branch of the inferior thyroid artery) pass upwards deep to the lower border of the inferior constrictor.

The **oesophagus** is the downward continuation of the pharynx. In the neck, it lies behind the trachea on the vertebral bodies. Laterally are the thyroid gland and carotid sheath and the thoracic duct lies behind its left border. The oesophagus enters the thorax somewhat to the left of the midline.

The region between the styloid process behind and the tongue and hyoid bone in front should now be cleaned. Identify the muscles which pass from the styloid process to the tongue, hyoid bone and pharynx, and the stylohyoid ligament which passes to the lesser horn of the hyoid bone. The glossopharyngeal nerve lies behind the stylopharyngeus muscle, and the digastric and stylohyoid muscles are superficial to all the other structures (Figs. 16 and 20). With bone forceps cut through the base of the styloid process and turn it forwards, taking care to preserve the glossopharyngeal nerve. Clean the structures which have been named above.

Cut through the posterior median raphe of the pharynx vertically and examine the inside of the pharynx (Fig. 50). The anterior wall of the pharynx presents from above downwards the posterior apertures of the nasal cavity (the choanae), the soft palate, the oral cavity, the base of the tongue, the **epiglottis,** the opening of the larynx and the posterior surface of the larynx. The posterior nasal apertures are separated from each other by the vertical posterior edge of the **vomer.** Behind the posterior nasal openings and just above the level of the palate, is the opening of the auditory tube with the **tubal elevation** formed by the projection of the cartilage

129

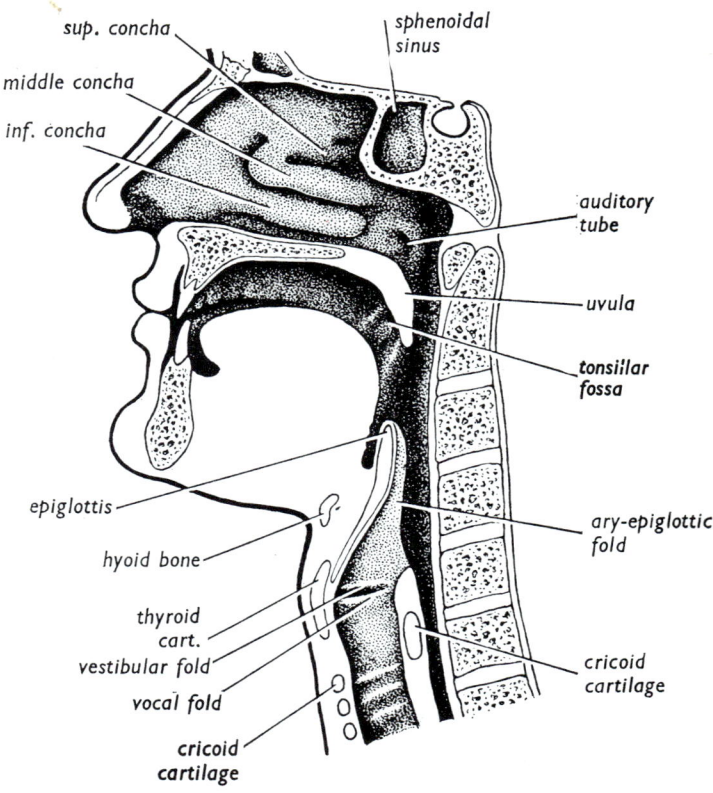

FIG. 50

Sagittal section through the head showing some of the relations of the pharynx and larynx.

of the tube on its superior and posterior sides. Passing from each side of the palate downwards to the tongue in front and to the pharynx behind are two folds of mucous membrane called the **arches of the fauces.** Each fold contains a muscle, and the **palatine tonsil** lies between them on each side of the base of the tongue. The faucial isthmus, formed by the arches, separates the mouth from the pharynx.

Re-examine the surface of both the tongue and the floor of the mouth (page 58). Dissect the mucous membrane from the superior surface of the soft palate and from the adjoining lateral wall of the pharynx. Identify the levator veli palatini and the tendon of tensor veli palatini, already seen from its lateral aspect. The tendon of the tensor is seen to pass round the hamulus of the medial pterygoid plate and then spread out in a thin sheet forming the **palatine aponeurosis** which is attached anteriorly to the posterior border of the hard palate. The fibres of the levator are more medial and are also attached to this aponeurosis.

Pull the soft palate upwards and fix it with a hook. Dissect the mucous membrane from its inferior surface and from the arches of the fauces. The anterior arch contains the **palatoglossus muscle** and the posterior arch the **palatopharyngeus muscle.** The isthmus of the fauces is bounded by the palate above, the palatoglossal arches laterally and the tongue below. The fibres of the palatopharyngeus intermingle with those of the stylopharyngeus and also with muscle fibres coming from the cartilaginous part of the auditory tube (the salpingopharyngeus muscle). The palatoglossus is attached to the inferior surface of the palatine aponeurosis and runs to the tongue where it becomes continuous with the transverse muscle of that organ (Fig. 20). Remove any tonsillar remains and expose the medial surface of the superior constrictor muscle. Note that the mucous membrane of the epiglottis is very adherent to the cartilage. Surrounding the opening of the larynx are the ary-epiglottic folds; they will be dissected later with the larynx and the tissues on the back of the cricoid cartilage.

STRUCTURAL DETAILS

The muscles and ligaments (Fig. 49)

The **superior constrictor muscle** is attached in front to the hamulus of the medial pterygoid plate and to a fibrous raphe passing from the hamulus to the posterior part of the mylohyoid line on the mandible. Attached to this **pterygomandibular raphe** are the superior constrictor behind and the buccinator muscle in front. The lowest part of the constrictor muscle is attached to the mylohyoid line. The fibres pass backwards and upwards to the posterior median raphe and the pharyngeal tubercle of the occipital bone. The lower fibres are covered laterally by the middle constrictor.

The **middle constrictor muscle** is attached to the lower part of the stylohyoid ligament and to the lesser and greater horns of the hyoid bone. The muscle fibres pass backwards and upwards to the median raphe and the lower fibres are overlapped laterally by the inferior constrictor.

The **inferior constrictor muscle** is attached to the oblique line on the lamina of the thyroid cartilage, to the lateral aspect of the cricoid cartilage and to the fascia covering the cricothyroid muscle. Its upper fibres pass backwards and upwards, overlapping the middle constrictor; its lower fibres, surrounding the narrowest part of the pharynx, merge with the circular muscle of the oesophagus. The constrictor muscles are supplied by branches from the pharyngeal plexus whose motor nerve is the pharyngeal branch of the vagus.

The **tensor veli palatini muscle** arises from the scaphoid fossa between the pterygoid plates, forms a tendon which passes round the hamulus and spreads out in the fibrous lamina of the soft palate. It is supplied by the mandibular division of the trigeminal nerve, the fibres passing without interruption through the otic ganglion (Fig. 32).

The **levator veli palatini muscle** is attached above to the rough inferior aspect of the apex of the petrous part of the temporal bone and passes down into the superior aspect of the soft palate. The muscle is supplied from the pharyngeal plexus of nerves. Between the tensor and levator muscles lies the **auditory tube.** The tensor is attached to the lateral aspect of the cartilaginous part of the tube and the levator to its medial aspect.

The **pharyngobasilar fascia** lies between the upper edge of the superior constrictor and the base of the skull, and is the thickened upper part of the fascia lying between the mucous membrane and the constrictor muscles.

The **styloglossus muscle** is attached to the styloid process and passes forwards to the side of the tongue. (Its nerve supply is from the hypoglossal nerve.) The **stylohyoid muscle** is attached to the styloid process and to the hyoid bone, where the muscle splits to enclose the tendon of the digastric muscle. (Its nerve supply is from the facial nerve.) The **stylopharyngeus muscle** is attached above to the styloid process and passes downwards between the superior and middle constrictor muscles to blend with the inner surface of the pharyngeal muscles. Some of its fibres are attached to the posterior border of the thyroid cartilage. (Its nerve supply is from the glossopharyngeal nerve.) The posterior belly of the **digastric muscle** is inferior and lateral to the stylohyoid muscle. It is attached to the mastoid notch medial to the mastoid process. The anterior belly is attached to the inner surface of the mandible near the midline. The two bellies unite in a tendon which is attached to the hyoid bone by a fascial sling and the stylohyoid muscle. (The posterior belly is supplied by the facial nerve and the anterior belly by the mylohyoid branch of the inferior alveolar nerve.)

The **stylohyoid ligament** is continuous with the tip of the styloid process above and with the lesser horn of the hyoid bone below.

The auditory tube

The auditory (Eustachian) tube joins the cavity of the middle ear to the cavity of the pharynx. The tube is bony in its lateral part and cartilaginous in its medial part. The mucous membrane of the tube may be involved in inflammation of the pharynx. As a result, the tube may become blocked and the air in the middle ear absorbed so that hearing is impaired. Swallowing opens the auditory tube and equalises the pressure on both sides of the ear drum. Sudden changes in atmospheric pressure due to changes in altitude can often be detected in the ears. In a child, the tube is more horizontal than in the adult in whom it passes upwards as well as laterally and backwards. Infection spreads more frequently from the pharynx to the middle ear in a child than in an adult.

The soft palate

The soft palate, with its dependent **uvula** in the middle of its posterior border, has respiratory mucous membrane on its superior and oral mucous membrane on its inferior surface. It is very mobile in the living and is elevated and made tense during swallowing to help close off the nasopharynx from the oropharynx. The tensor and levator veli palatini enter the palate above and the palatoglossus and palatopharyngeus leave it below.

The palatoglossal folds pass into the sulcus terminalis on the dorsum of the tongue. The palatopharyngeal folds pass down

133

behind the tonsil. The muscle fibres of the palatopharyngeus merge with the middle constrictor and some are attached to the thyroid cartilage.

The tonsil

The palatine tonsil is usually very shrunken or absent in cadavers but is easily seen and examined, if present, in living subjects. The medial surface in the living subject is red and pitted by the openings of the **tonsillar crypts**. The tonsil is firmly attached at its inferior pole where most of the blood vessels enter and in front where it lies behind the palato-glossal fold. There is usually a well marked cleft between the superior pole of the tonsil and the upper wall of the tonsillar fossa. Laterally a fibrous capsule separates the tonsil from the superior constrictor (Fig. 51). If the facial artery arches upwards high enough, lateral to the superior constrictor, it becomes a lateral relation of the tonsil (Fig. 49).

The main artery to the tonsil comes from the facial but it also receives branches from the ascending pharyngeal and lingual arteries. The arteries reach the tonsil by passing over the upper border of and through the superior constrictor. A large vein passes laterally through this muscle to join the pterygoid plexus of veins.

The tonsil consists mainly of lymphoid tissue. Accumulations of lymphoid tissue are also found in the posterior third of the tongue (the **lingual tonsil)** and beneath the mucous membrane of the back of the roof of the pharynx and adjacent posterior wall (the **pharyngeal tonsil** or **adenoid**). Here the lymphoid tissue is in close relation to the occipital bone and the anterior arch of the atlas and completes a ring of lymphoid tissue round the upper part of the pharynx.

The epiglottis

The epiglottis is " leaf-shaped " and has a smooth anterior surface which projects upwards behind the base of the tongue. When the upper edge of the epiglottis is pulled backwards the median glosso-epiglottic fold is seen in the midline with a hollow on each side called the **vallecula.** The lateral limit of this hollow is the wall of the pharynx raised to form the lateral glosso-epiglottic fold. Passing posteriorly and downwards from the sides of the epi-

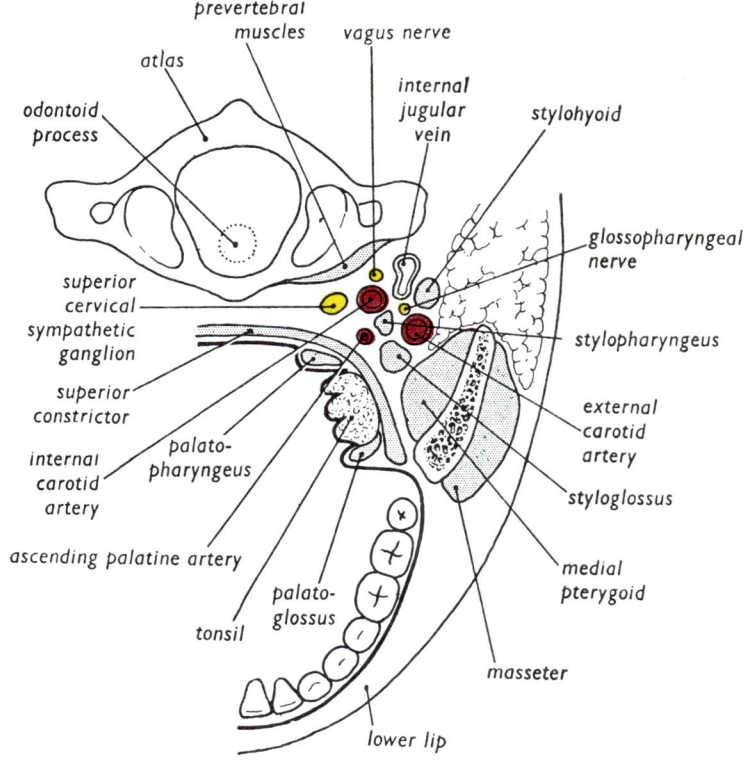

FIG. 51

Diagram of an oblique section through the palatine tonsil.

glottis are the **ary-epiglottic folds** containing muscle fibres from the laryngeal musculature. The middle of the posterior surface of the epiglottis bulges backwards to form the **tubercle of the epiglottis.** Below the ary-epiglottic folds the anterior wall of the pharynx is formed by the posterior aspect of the larynx. This consists of the expanded posterior part of the cricoid cartilage with the **arytenoid cartilages** embedded in muscle on its upper border. The posterior borders of the laminae of the thyroid cartilage lie laterally. Below this, the pharynx narrows to become continuous with the oesophagus. Lying lateral to the ary-epiglottic folds and medial to the thyroid laminae are the **piriform**

135

fossae. The close relation of the internal laryngeal nerve to the anterior border of the piriform fossa may be demonstrated by pulling on the nerve where it pierces the thyrohyoid membrane.

The pharyngeal plexus of nerves

The nerve supply of the pharynx and palate is derived mainly from this plexus which lies on the superficial aspect of the middle constrictor muscle. The plexus receives (1) a branch from the glossopharyngeal nerve which is sensory, (2) the pharyngeal branch of the vagus which is motor (most of this branch is derived from the cranial part of the accessory nerve) and (3) a branch from the cervical sympathetic trunk. All the pharyngeal and palatal muscles are supplied by this plexus except the stylopharyngeus muscle, which is supplied directly from the glossopharyngeal nerve, and the tensor veli palatini which is supplied by the mandibular division of the trigeminal nerve through its medial pterygoid branch. Sensory fibres have their cell bodies in the ganglia of the glossopharyngeal and vagus nerves and their central processes end in the medullary nucleus of the vagus (general sensation) and the nucleus of the tractus solitarius (taste sensation). The motor fibres innervating the pharynx, palate and larynx have their cell bodies in the nucleus ambiguus of the medulla oblongata and leave the hindbrain mainly in the cranial part of the accessory nerve. These fibres join the vagus nerve and are distributed through its pharyngeal and laryngeal branches. A few fibres pass into the glossopharyngeal nerve.

The **glossopharyngeal nerve** (Fig. 20) leaves the skull through the jugular foramen and lies deep to the styloid process and the structures attached to it. The nerve turns round the posterior border of the stylopharyngeus muscle and passes forwards superficial to it. It then lies deep to the hyoglossus muscle and ends in the mucous membrane of the posterior third of the tongue. It supplies a motor branch (from the nucleus ambiguus) to the stylopharyngeus, and sensory fibres to the mucous membrane of the pharynx, the tonsil and, to some extent, the soft palate (through its branch to the pharyngeal plexus), and the posterior third of the tongue (including taste fibres). It also gives a sensory branch to the carotid sinus and the carotid body, and the parasympathetic (secretomotor) fibres to the parotid gland which synapse in the otic ganglion (**Fig. 32**).

136

FUNCTIONAL ASPECTS

Deglutition

Swallowing a bolus of food placed on the anterior part of the tongue begins as a voluntary movement, but as the pharyngeal musculature and oesophagus take over control of the bolus, the movements become involuntary. Firstly, the bolus is pushed on to the posterior part of the tongue by elevation of the tip and pressure by the tongue against the hard palate from before backwards. Before the bolus leaves the back of the tongue, the nasopharynx is closed off from the oropharynx by elevation of the soft palate and its approximation to the upper part of the pharyngeal wall. This is produced by contraction of the levator and tensor veli palatini and by the sphincter action of some fibres of the superior constrictor producing a ridge (of Passavant) on the lateral and posterior walls of the pharynx. Secondly, the bolus is pushed through the oropharyngeal isthmus by contraction of the palatoglossus and by elevation of the hyoid bone and the base of the tongue. This elevation is brought about by contraction of the geniohyoid, mylohyoid, stylohyoid, styloglossus and digastric muscles. During these movements, the mandible is held in position by its elevators. The larynx is shut off from the pharynx by elevation of the larynx towards the epiglottis and by the partial closure of the laryngeal opening due to contraction of the muscle in the ary-epiglottic folds. The larynx is raised by the palatopharyngeus, salpingopharyngeus, stylopharyngeus and thyrohyoid muscles. Thirdly, the bolus is passed on by successive contractions of the superior, middle and inferior constrictors into the oesophagus. After the bolus has passed beyond the laryngeal opening the hyoid, lingual and palatine muscles relax. This allows the larynx to return to its lower position and, with the opening of the sphincters between the nasopharynx and oropharynx, air can be inhaled without the danger of food being inhaled at the same time. When the lips and jaws are parted, swallowing is difficult.

The danger of inhalation of food is minimised by two further factors. When the larynx is raised, the pharyngeal mucous membrane is also raised and when muscle relaxation occurs, the larynx descends more quickly than the mucous membrane which is left covering the opening of the larynx. In addition when food reaches the back of the tongue, the more solid matter passes over the tip of the epiglottis even bending it backwards and liquid food passes down on either side of the epiglottis. The result is, normally, neither solids nor liquids enter the larynx. There is radiographic evidence for all these mechanisms. (In some animals the epiglottis fits accurately against the palate and so the food column is split and passes on either side of the opening of the larynx.)

THE LARYNX

INTRODUCTION

THE larynx opens into the pharynx above at the laryngeal opening and is continuous below with the trachea. It lies anterior to the lower part of the pharynx and in the adult is opposite the 3rd, 4th, 5th and 6th cervical vertebrae. In infants, however, the upper edge of the epiglottis is as high as the plane between the 1st and 2nd cervical vertebrae. In the male, the antero-posterior dimension of the larynx lengthens considerably at puberty whereas in the female it shows little increase. Look into the larynx in the cadaver through the laryngeal opening and identify the **vestibular folds** superiorly and the **vocal folds** inferiorly. In the living subject these structures may be seen by means of a laryngeal mirror.

The upper opening of the larynx lies at a level between the hyoid bone and the upper part of the thyroid cartilage. The thyroid cartilage forms in front the prominence of the " Adam's apple," and its two wing-like laminae are largely covered by the thyroid gland and infrahyoid muscles. There is a synovial joint between the lower posterior end of each thyroid lamina and the posterolateral surface of the arch of the cricoid. (The cricoid cartilage has been likened to a signet ring.) On the lateral part of the upper border of the broad lamina of the cricoid are the two arytenoid cartilages. The vocal folds pass from the inner aspect of the thyroid cartilage near the midline backwards to the arytenoid cartilages. Below the cricoid cartilage are the incomplete cartilaginous rings of the trachea.

DISSECTION

(This dissection may be omitted.)

Make a median vertical incision through the mucous membrane of the pharynx which lies on the posterior surface of the larynx and dissect out the laryngeal muscles on one side. While removing the mucous membrane laterally look for the recurrent laryngeal nerve as it passes posterior to the cricothyroid articulation and divides into branches that supply all the muscles of the larynx except the cricothyroid. Between the posterior surfaces of the two arytenoid cartilages some muscle fibres pass transversely from one arytenoid to the other (the **transverse arytenoid muscle**) and others form an " X " on the back of the arytenoids (the **oblique arytenoid muscle**).

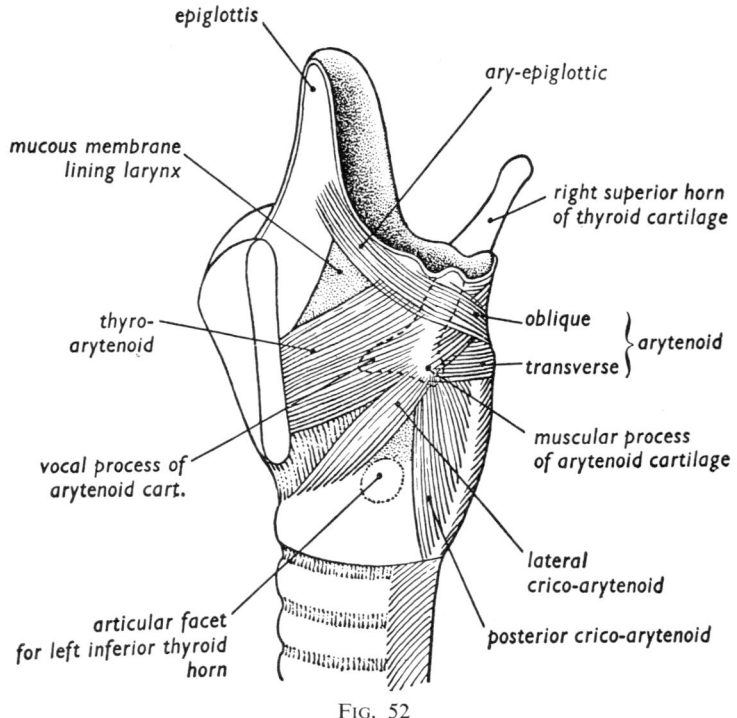

FIG. 52

Drawing of the muscles attached to the left arytenoid cartilage.

Some of these oblique fibres can be traced upwards into the ary-epiglottic folds, where they form part of the ary-epiglottic muscle. The **posterior crico-arytenoid muscle** is attached to one half of the posterior surface of the lamina of the cricoid cartilage and passes upwards and laterally to the posterior aspect of the muscular (lateral) process of the arytenoid cartilage. Dissect between the cricoid and thyroid cartilages to expose the **lateral crico-arytenoid muscle,** attached to the upper border of the arch of the cricoid cartilage and passing upwards and backwards to the anterior aspect of the muscular process of the arytenoid cartilage.

On the outside of the larynx identify the **cricothyroid muscle** whose fibres pass backward from the arch of the cricoid to the lower border and inferior horn of the thyroid cartilage. Trace the external laryngeal nerve to the cricothyroid muscle. Remove the left lamina of the thyroid cartilage together with the cricothyroid muscle by dividing the cartilage vertically to the left of the midline near the anterior border and dissecting it away from the

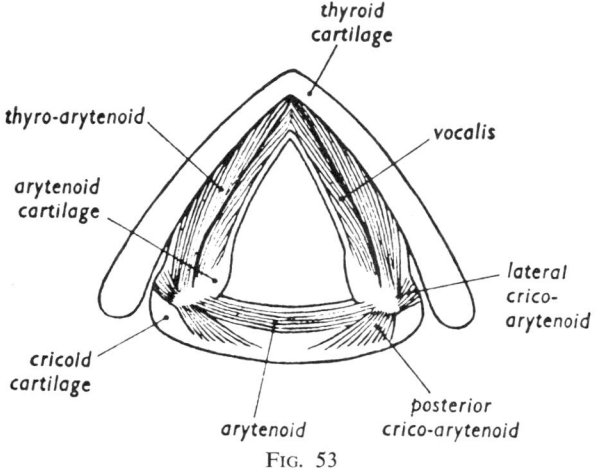

FIG. 53

The muscles attached to the arytenoid cartilages.

underlying muscle and mucous membrane (Fig. 52). The lateral crico-arytenoid muscle is seen passing backwards from the cricoid arch to the arytenoid cartilage. Above this the broad sheet of the thyro-arytenoid muscle will be seen. The **thyro-arytenoid muscle** is attached in front to the internal aspect of the laryngeal prominence below the attachment of the epiglottis, and behind to the antero-lateral surface of the arytenoid cartilage. Some of the uppermost fibres sweep superiorly into the ary-epiglottic folds (the **thyro-epiglottic muscle).** On the medial side of the muscle there are fibres which are lateral to and attached to the vocal ligament within the vocal fold. They are called the **vocalis muscle** and are attached posteriorly to the vocal process of the arytenoid cartilage (Fig. 53).

Divide the posterior aspect of the larynx by a median vertical incision through muscle and cartilage and examine its interior. Two folds of mucous membrane are seen on each side. The upper **(vestibular)** fold forms the lower boundary of the **vestibule** of the larynx lying below the ary-epiglottic folds. (The vestibular fold is also called the false vocal cord.) The lower **(vocal)** fold (or true vocal cord) is covered by stratified squamous epithelium and the rest of the interior of the larynx by respiratory ciliated mucous

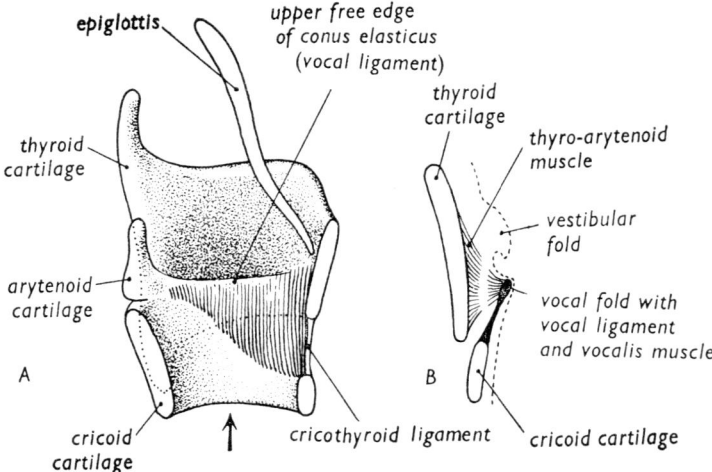

FIG. 54

A. The relation of the conus elasticus to the cartilages on the left side is indicated.
B. A coronal section of the larynx cut at the point indicated by the arrow.

columnar epithelium. In the vocal fold are the vocal ligament and vocalis muscle. The **rima glottidis** is formed by the edges of the vocal folds and the medial surfaces of the arytenoid cartilages. Between the vocal and vestibular folds is the **sinus** of the larynx leading into a recess, the **laryngeal saccule,** which extends a short distance upwards, lateral to the vestibular folds (Fig. 54B).

The terminal branches of the internal laryngeal nerve supply sensory branches to the laryngeal mucous membrane down to the level of the vocal folds. The branches of the recurrent laryngeal nerve can also be traced from just behind the cricothyroid articulation. This nerve supplies all the muscles of the larynx except the cricothyroid muscle and is sensory to the mucous membrane below the vocal folds.

The **thyrohyoid membrane** passes upwards from the upper border of the thyroid cartilage and then behind the hyoid bone, and is attached to the superior border of its body. The lateral borders of the membrane are thickened to form the **lateral thyrohyoid ligaments.**

Examine the space between the cricoid and thyroid cartilages. It is occupied by a membrane that is attached below to the whole of the arch of the cricoid. Its upper attachments are more complicated (Fig. 54). Anteriorly the membrane is attached to the thyroid cartilage in the midline. Posteriorly the membrane passes from the cricoid cartilage to the vocal process of the arytenoid cartilage. Between the thyroid cartilage in front and the arytenoid cartilage behind, the membrane passes upwards deep to the thyroid cartilage and has a thickened, free, upper border (the **vocal ligament)** lying in the vocal fold. The part of the membrane lying deep to the thyroid lamina is called the **conus elasticus** and is separated from the cartilage by the thyro-arytenoid muscle. Make a vertical incision through one of the vocal folds and identify the conus elasticus and the vocal ligament medially, and the vocalis and the thyro-arytenoid muscles laterally.

Open the cricothyroid and crico-arytenoid joints. These are synovial joints with a lax capsule allowing gliding as well as rotatory movements.

The **trachea** is the downward continuation of the larynx and is kept patent by C-shaped cartilages in its walls. They are deficient posteriorly where the fibromuscular wall of the trachea is flattened and in contact with the oesophagus. Lateral to the trachea on each side is the common carotid artery and the lobe of the thyroid gland. Between the trachea and oesophagus laterally is the recurrent laryngeal nerve. The isthmus of the thyroid gland is anterior to the 2nd, 3rd and 4th rings of the trachea, and below this, in the living, rings of the the trachea can be felt deep to the skin in the midline. As it enters the thorax the trachea passes somewhat backwards behind the manubrium sterni.

STRUCTURAL DETAILS

The laryngeal cartilages

The **thyroid cartilage** consists of two flat quadrilateral **laminae** joined together in front to form the **laryngeal prominence.** Above the prominence, the laminae are separated by the **thyroid notch** which is easily palpable in the living subject. The angle formed by the laminae shows considerable sex difference, the average in the male being about 90° and in the female 120°.

The posterior border of the lamina projects upwards and downwards beyond its upper and lower borders as the **superior** and **inferior horns.** The superior horn is connected to the tip of the greater horn of the hyoid bone by the lateral thyrohyoid ligament. The inferior horn articulates with the posterolateral aspect of the cricoid cartilage. The lateral surface of the lamina has an **oblique line** passing downwards and forwards from the root of the superior horn to the inferior border.

The **cricoid cartilage** forms part of the skeleton of the larynx below the thyroid cartilage. It is shaped like a signet ring, with a broad **lamina** posteriorly and a narrow **arch** completing the ring laterally and anteriorly. The upper border of the lamina has a facet on each side of the midline for articulation with the base of an arytenoid cartilage. There is a facet on the posterolateral aspect of the arch for articulation with an inferior horn of the thyroid cartilage. All these joints are synovial.

The **arytenoid cartilages,** right and left, are three-sided pyramids articulating by their bases with the lateral part of the upper border of the cricoid lamina. The apex of each extends upwards and medially into the ary-epiglottic fold. The base of the arytenoid cartilage projects forwards as the **vocal process** for the attachment of the vocal fold, and laterally as the **muscular process** for the attachments of the lateral and the posterior crico-arytenoid muscles. Attached to the posterior surface of these cartilages is the arytenoid muscle and to their lateral surface the thyro-arytenoid muscle.

The **epiglottis** is a leaf-shaped plate of cartilage lying behind the tongue and hyoid bone. Its narrow inferior end is attached to the posterior surface of the thyroid cartilage below the notch, and its broad superior end forms a free projection behind the tongue.

Its lateral borders are in the ary-epiglottic folds. On its posterior surface are deep pits containing mucous glands.

Structure of the laryngeal cartilages

The thyroid, cricoid and most of the arytenoid cartilages are composed of hyaline cartilage. They tend to calcify in adult life and may then be visible on X-ray films.

The vocal process and the apex of the arytenoid cartilages, and the epiglottis are composed of elastic cartilage and do not become calcified.

FUNCTIONAL ASPECTS

The actions of the laryngeal muscles

The thyroid and cricoid cartilages can rotate round a transverse axis through the cricothyroid joints and can also glide on each other. The arytenoid cartilages can be moved on the cricoid cartilage either by gliding medially or laterally, or by rotating so that the vocal processes and consequently the vocal folds approach or diverge from one another.

The **cricothyroid muscles** move the cricoid cartilage upwards and backwards on the inferior horns of the thyroid cartilage. It also pulls the whole thyroid cartilage forwards. The effect of this is to lengthen (and tense) the vocal folds. The **thyro-arytenoids** pull the arytenoid cartilages towards the thyroid so shortening (and relaxing) the vocal folds. They also adduct the vocal folds. The **vocalis muscle** increases the tension within the vocal fold. The **lateral crico-arytenoid** adducts the vocal fold, and the **posterior crico-arytenoid** abducts the vocal fold. The **arytenoid muscle** pulls the arytenoid cartilages together and narrows the cleft between the cords. The **ary-epiglottic** and **thyro-epiglottic muscles** approximate the ary-epiglottic folds to one another and help to close the laryngeal opening.

The larynx in swallowing, exercise and speech

The function of the larynx is mainly to keep food out of the respiratory tracts by providing a sphincter mechanism for closing off the upper end of the respiratory tract. The chest is fixed in strenuous muscular effort during which the sphincter may be closed. A man with a permanent artificial opening in the trachea, *e.g.* after removal of the larynx, finds hard manual work very difficult. The principal sphincter muscles are the arytenoideus, thyro-arytenoid, and lateral crico-arytenoid (adductor of the vocal folds) aided by the ary-epiglottic muscles. The dilator action is produced by the pressure of air on the relaxed muscles and by the posterior crico-arytenoid muscle (abductor of the vocal folds). During quiet breathing the folds are in a neutral position (about midway between full adduction and full abduction) and relaxed, although there is slight abduction on inspiration.

In whispering, the vocal folds are separated and relaxed. In voiced sounds, the vocal folds are adducted. The expiratory air stream is pushed past the folds and interrupted at various frequencies. This variation may be the chief factor in changing the pitch of the voice but the vibrating length, the tension and the thickness of the vocal folds may play a part. For example, the male voice is deeper than the female due to the male vocal folds being longer. In a falsetto voice only the anterior halves of the folds are functioning and are very thin.

The vibration in the larynx is transmitted to the trachea and lungs, and the pharynx, mouth and nose. This gives quality to the note. The loudness of the note depends on the volume of air being expired. In speech, the laryngeal air stream is modified by the tongue, palate, teeth, lips and cheeks so that different vowels and consonants are produced. Recognisable speech depends largely on special parts of the brain.

CHAPTER 15

THE NASAL CAVITY

INTRODUCTION

M OST students fail to appreciate the extent of the nasal cavity since their attention is usually concentrated on the external nose. Examination of the hard palate on the base of the skull gives a much better idea of the anteroposterior and transverse extent of the floor of the nasal cavity. The **septum** divides the cavity into right and left halves. The **external nose** has a partly bony and partly cartilaginous skeleton. The shape of the nose, its width and the direction of the nostrils vary in different people. The nasal bones, meeting in the midline and articulating with the frontal processes of the maxilla, form the upper and lateral parts of the external nose. The remainder consists of several cartilages attached to the nasal bones and to the edges of the maxillae which surround the large opening seen in the middle of the front of the skull. The skin is freely movable on the bony skeleton but is more firmly bound down to the cartilages which surround the anterior apertures (the nostrils or nares). A separate cartilage forms the upper lateral wall of the external nose. The lower lateral wall is called the **ala,** and its cartilage also forms part of the septum in the midline between the nares. Verify this on your own nose on which it is also possible to distinguish how much of the external nose is cartilage and how much bone.

The surface area of the lining of the nasal cavities is increased by folds (the **conchae**) on the lateral walls and by a series of diverticula (the **sinuses**). When air is breathed in through the nose, it is warmed and moistened by contact with the lining of the nasal cavity before it passes downwards to the lungs. The most important function of the nasal cavity is respiratory. In addition, the sense of smell is located in the lining of its roof, and in speech the nasal cavity is used to produce the nasal consonants (m, n, ng).

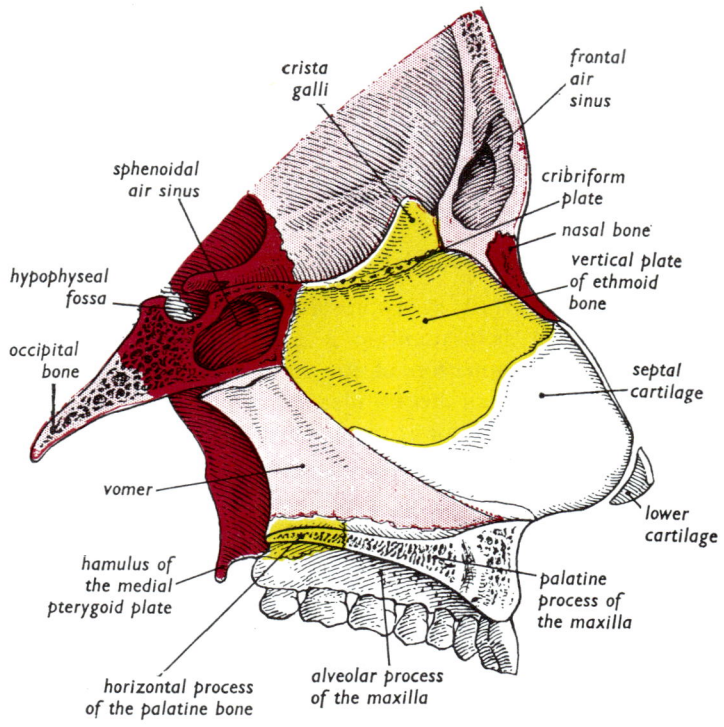

FIG. 55

A paramedian section of the front of the skull showing the septum and the bones forming the roof and floor of the right nasal cavity.

DISSECTION

Make a sagittal saw-cut in the anterior half of the skull slightly to one side of the midline (use a knife to cut through cartilage and the soft palate) and examine the nasal septum on the half of the skull on which it remains. Just within the nostril on the septum there is a depression which is the medial wall of the **vestibule** and on it are some stiff hairs called **vibrissae.** Remove the mucous membrane and examine the different parts of the septum. It is rarely in the midline and may be so deviated to one side that it occludes one nasal cavity. The nasal **septum** consists mainly of parts of two bones and the septal cartilage (Fig. 55). The vomer forms

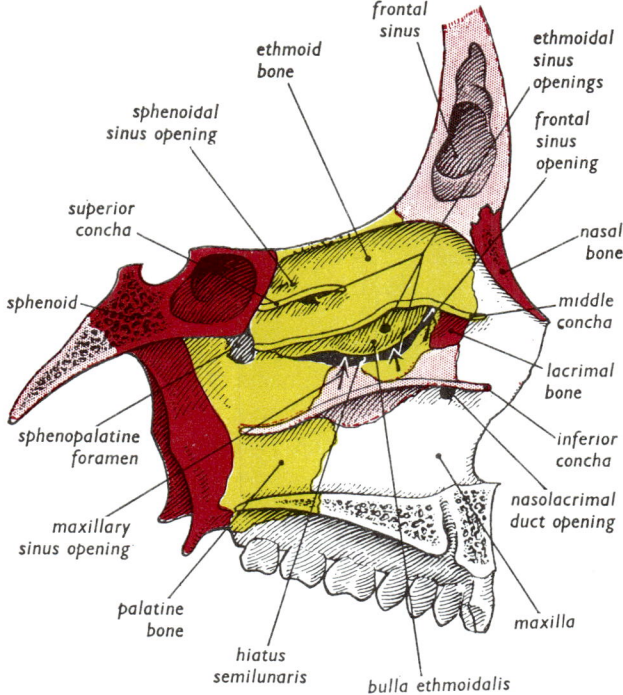

FIG. 56

The skeleton of the lateral wall of the left nasal cavity. Most of the conchae have been cut away.

the postero-inferior part, the perpendicular plate of the ethmoid forms the posterosuperior part, and the septal cartilage forms the anterior part. The two bones have irregular shapes and the septal cartilage fits into the angle between them. The septal cartilage projects forwards and its anterior border helps to give the nose its characteristic outline.

Find on the septum the nasopalatine nerve running obliquely down towards the incisive canal, through which it passes to the oral surface of the palate. Bundles of olfactory nerves may be found passing through the cribriform plate in the roof. Remove the septum, if necessary, and examine the lateral wall of the cavity. Immediately above the nostril is the lateral wall of the vestibule on which vibrissae can be seen. The most striking feature of the lateral wall is the irregularity due to the projection of the three **conchae** (Figs. 50 and 56). The **inferior concha,** the largest of the three, is a separate bone. Like the others, it projects medially and then downwards. The space between the lateral wall and the inferior concha is called the **inferior meatus.** Remove a small portion of the anterior part of the inferior concha and find on the lateral nasal wall the opening of the **nasolacrimal duct** which conveys the tears from the lacrimal sac into the nose. Look for the pharyngeal opening of the **auditory tube,** on a level with the inferior meatus but posterior to the nasal cavity.

The **middle concha,** nearly as big as the inferior, is part of the ethmoid bone. Between the attachments of the middle and inferior conchae is the **middle meatus** of the nose. To examine this meatus, remove most of the middle concha. Identify a rounded projection, the **bulla ethmoidalis.** It is limited inferiorly by a groove, the **hiatus semilunaris.** In the front of the hiatus find the lower opening of the **frontonasal duct.** Try to pass a probe upwards along the duct into the frontal sinus. The lower open part of the frontonasal duct is often called the infundibulum and on to this open the anterior ethmoidal air cells. The bulla ethmoidalis is formed by the underlying middle ethmoidal air cells which open on to the surface of the bulla (Fig. 56). In the hiatus semilunaris posteriorly is the opening of the maxillary sinus. Compare the size of this maxillary opening with that on the medial wall of a disarticulated maxilla. The inferior concha, lacrimal and palatine bones and the bulla ethmoidalis close most of the opening in the living, and note that the level of the opening is higher than the level of the floor of the sinus. To determine the extent of the maxillary sinus, remove carefully the inferior concha. Examine the floor of the sinus and note its relation to the alveolar process and the roots of the teeth.

The **superior concha,** the smallest, is also part of the ethmoid bone. It is on the upper and posterior part of the lateral wall of the nose. Between the attached edges of the superior and middle conchae is the **superior meatus** into which the posterior ethmoidal air cells open by one or more apertures. Above the superior concha is a space called the **spheno-ethmoidal recess.** Look for the opening of the sphenoidal sinus in the posterior part of this recess. The opening is on the front wall of the sinus. The highest part of the roof of the nose is formed by the cribriform plate of the ethmoid. Followed backwards, the roof slopes downwards along the anterior and inferior surfaces of the body of the sphenoid. Followed forwards, it slopes downwards along the inferior surfaces of the frontal and nasal bones. The floor is formed by the upper surface of the hard and soft palates and is wider from side to side than the roof. The height of the nasal cavity is about 5 cm and the anteroposterior length is about 7 cm.

Examine the mucous membrane lining the nasal cavity and note that it is firmly adherent to the underlying periosteum, forming a mucoperiosteum. It is thicker and more vascular over the conchae, especially the inferior. The openings of the ducts of mucous glands can be seen. The mucoperiosteum of the nasal cavity is continuous with that lining the paranasal sinuses. Find the **sphenopalatine foramen** just above the posterior end of the middle concha by carefully removing the mucoperiosteum (Fig. 56). The **nasal** and **nasopalatine** branches of the maxillary nerve enter the nasal cavity through this foramen. With them is the **sphenopalatine** branch of the maxillary artery. Find the cut end of the **nasopalatine nerve.** The nerve crosses the roof of the nasal cavity and passes downwards and forwards on the septum to the floor. It then passes through the incisive canal and supplies branches to the oral surface of the palate. The **nasal nerves** are distributed to the posterosuperior parts of the nasal cavity.

In a groove on the deep surface of the nasal bone is the **anterior ethmoidal nerve,** a branch of the nasociliary part of the ophthalmic nerve. It can be traced upwards to the cribriform plate of the ethmoid, at the side of which it enters the skull. This nerve gives branches to both the medial and lateral walls of the cavity and an external nasal branch to the skin of the side of the external nose. Accompanying the ethmoidal nerve is the **anterior ethmoidal artery,** a branch of the ophthalmic artery.

The main arteries to the walls of the nasal cavity are the sphenopalatine, the descending palatine and the anterior ethmoidal. Branches of these join with vessels from the upper lip to form an anastomosis on the septal wall just above the vestibule; this is frequently the site of nose bleeding.

151

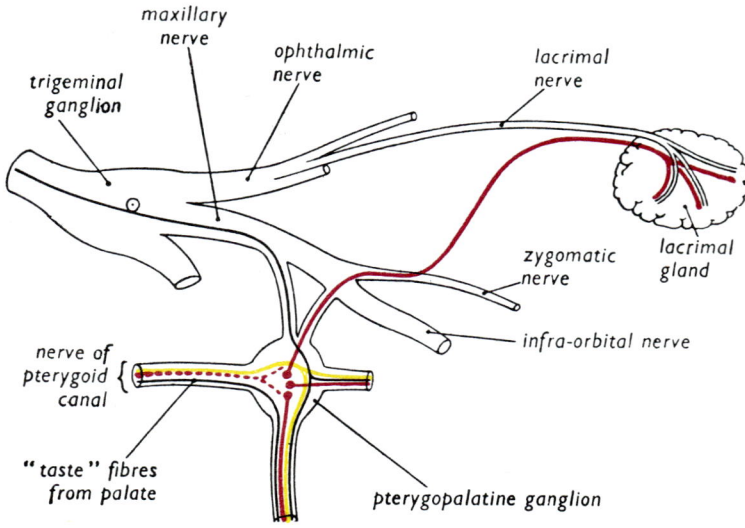

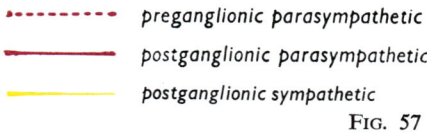

·–·–·–·–· preganglionic parasympathetic

⸻⸻⸻ postganglionic parasympathetic

⸻⸻⸻ postganglionic sympathetic

FIG. 57

Diagram of the connexions of the pterygopalatine ganglion. The pathway of the parasympathetic (secretomotor) fibres to the lacrimal gland is shown. Sensory and postganglionic sympathetic and parasympathetic branches to the nasal and palatine mucous membrane are indicated.

An attempt may now be made to dissect out the maxillary nerve, the maxillary artery and the pterygopalatine ganglion as they lie in the pterygopalatine fossa. Beginning at the sphenopalatine foramen and working downwards remove the perpendicular plate of the palatine bone with a chisel and open up the pterygopalatine fossa from its medial side.

The **greater palatine nerve** is a branch of the maxillary nerve. It passes downwards along the posterior edge of the nasal cavity within the bones, gives branches to the lateral wall of the nose and appears at the back of the hard palate through the greater palatine foramen.

152

In the upper part of the fossa find the **maxillary nerve** and the **pterygopalatine ganglion** attached to it. If traced backwards, the nerve is found to emerge from the foramen rotundum, and if traced forwards to pass through the inferior orbital fissure. Its terminal branch is the infra-orbital nerve, which has been seen on the face. The pterygopalatine ganglion contains nerve cell bodies whose axons are distributed to the lacrimal gland and to the mucous glands in the walls of the nose and the palate (Fig. 57). The preganglionic branches arise in the superior salivary nucleus of the medulla and travel in the facial nerve, then in its greater petrosal branch and through the pterygoid canal in the nerve of the pterygoid canal. These parasympathetic fibres produce secretomotor and vasodilator effects. The sympathetic fibres, derived from the carotid plexus, produce vasoconstriction. Many sensory branches of the maxillary nerve pass through the ganglion and are distributed to the pharynx, the palate, the nose and the orbit.

The **posterior superior alveolar** branches of the maxillary nerve may be seen entering small bony canals in that part of the maxilla which forms the anterior wall of the pterygopalatine fossa. The nerves run in the posterior and lateral walls of the maxillary sinus and give branches to the molar and premolar teeth. In the infra-orbital canal, the infra-orbital nerve gives off the **anterior superior alveolar** branch which has a long course in the anterior wall of the sinus and gives branches supplying the incisor and canine teeth, the floor of the nasal cavity and the lining of the sinus.

In the pterygopalatine fossa is the end of the **maxillary artery**. It gives off terminal branches to the nose (sphenopalatine), to the palate (greater palatine), to the upper jaw (infra-orbital) and to the pharynx (pharyngeal).

STRUCTURAL DETAILS

The bones of the nasal cavity

The **ethmoid bone** ossifies in the cartilage of the wall of the nasal capsule adjacent to the cranial cavity. It forms part of the floor of the anterior cranial fossa, the lateral walls and septum of the nasal cavity, and the medial walls of the orbits. It consists of a central **perpendicular plate,** and of two lateral masses

153

(ethmoidal labyrinths), united at their upper extremities by the thin horizontal **cribriform plate** pierced by holes for the olfactory nerve bundles. In addition, a median vertical crest, the **crista galli,** extends upwards into the cranial cavity from the cribriform plate and gives attachment to the anterior end of the falx cerebri. The perpendicular plate forms the upper part of the nasal septum (Fig. 55). Each ethmoidal labyrinth contains a number of thin-walled air cells so that the bone is light. The lateral wall of the labyrinth is thin and forms part of the medial wall of the orbit. The medial wall of the labyrinth forms part of the lateral wall of the nasal cavity. The **superior** and **middle nasal conchae** project medially and downwards into the nasal cavity from this medial surface. Below the middle concha the air cells form a bulge, the **bulla ethmoidalis** (Fig. 58).

The **inferior concha** is a thin plate of bone projecting medially and downwards from the lower part of the lateral wall of the nasal cavity (Fig. 58). Its lower free border lies about 1 cm above the floor of the nose. The concha reduces the size of the opening in the medial wall of the maxilla.

The **lacrimal** is a small bone in the anterior part of the medial wall of the orbit. Its lateral (orbital) surface has a groove anteriorly, which, combined with the adjacent groove on the frontal process of the maxilla, forms the fossa for the lacrimal sac. Its medial (nasal) surface forms part of the lateral wall of the nasal cavity.

The **vomer** is a thin plate of bone forming the posterior part of the nasal septum, including its free posterior border (Fig. 55). It articulates above with the body of the sphenoid and below with the hard palate.

The **nasal bones** form the skeleton of the bridge of the nose and articulate with each other in the midline, and laterally with the frontal process of the maxilla.

Each **palatine bone** (Fig. 56) consists of a **horizontal plate,** forming the posterior third of the hard palate, and a **perpendicular plate,** forming the lateral wall of the nose between the maxilla in front and the medial pterygoid plate behind. The horizontal and perpendicular plates unite at right angles, and from their junction the **tubercle** projects backwards into the gap between the lower part

154

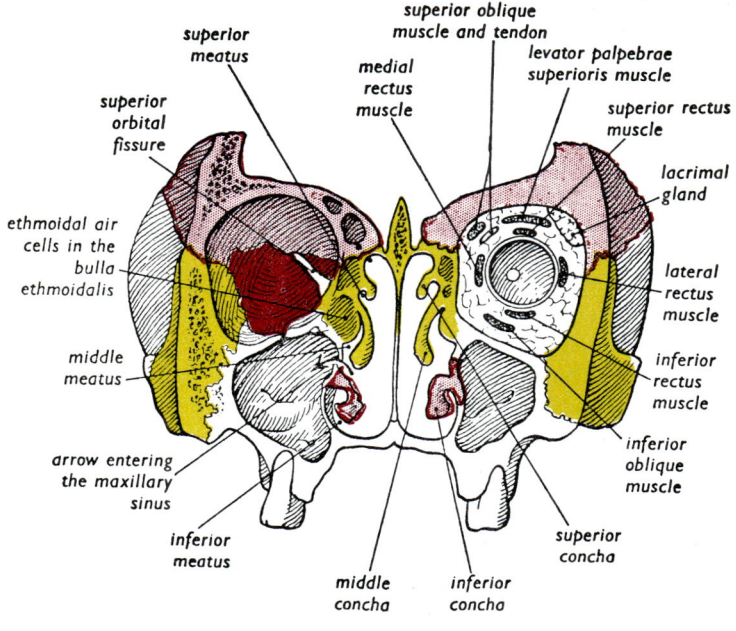

FIG. 58

Diagram of a coronal section through the orbits. Note the relations of the anterior cranial fossa, the orbits, the nasal cavities, the maxillary sinuses and the mouth to one another.

of the anterior margins of the medial and lateral pterygoid plates (Fig. 1). The upper end of the perpendicular plate divides into an orbital process laterally and a sphenoidal process medially. Between them is the sphenopalatine notch.

The **maxilla** is described on page 26.

The paranasal sinuses (Figs. 56 and 58)

These are in the frontal, ethmoidal, sphenoidal and **maxillary** bones. Their mucous membrane is continuous with that of the nasal cavity and their openings have already been identified. The sinuses are often the seat of disease and a knowledge of their relations is necessary to appreciate what other structures may be involved.

155

The **frontal sinuses** lie above the superciliary arches in the squamous part of the frontal bone separated from each other by a septum which is rarely in the midline. They are about 3 cm high, 2·5 cm wide and 2 cm deep. The midsagittal section of the skull usually opens one of the sinuses. Remove the anterior wall, if the sinus has not yet been opened. Confirm that the sinus opens into the middle meatus of the nose through the frontonasal duct. This duct is well placed for drainage of the sinus when the body is in the upright position, though drainage is difficult if the duct becomes blocked due to inflammation. The most important relations of the sinus are the anterior cranial fossa with the meninges and frontal lobe of the brain behind, the orbit below and lateral, and the nasal cavity below and medial.

The **ethmoidal labyrinth** consists of a large number of thin-walled air cells (Figs. 56 and 58), usually described as being arranged in three groups, anterior, middle and posterior. The former two open into the middle meatus and the latter into the superior meatus. Their most important relation is the orbit laterally from which they are separated by only a thin plate of bone. Infection in the orbit may be caused by infection of the ethmoidal air sinuses.

The **sphenoidal sinuses** are in the body of the sphenoid bone behind the upper part of the nasal cavities (Fig. 55). The septum separating them is rarely in the midline. They measure approximately 2 cm in each direction and open anteriorly into the spheno-ethmoidal recess. This opening is not always at the lower end of the anterior wall of the sinus so that drainage in the upright position may be inefficient. Drainage, however, is usually efficient in the prone position. The important relations of this sinus are the optic chiasma and the hypophysis cerebri above, and the cavernous venous sinus and internal carotid artery laterally. The relations of the sphenoidal sinus are best studied in a sagittal section of the skull (Figs. 55 and 56). *antrum of Highmore –*

The **maxillary sinus,** the largest of the sinuses, is the cavity in the maxillary bone and is pyramidal in shape (Fig. 58). Its base forms part of the lateral wall of the nose, and its apex extends laterally into the zygomatic process of the maxilla. It has an upper surface which forms the floor of the orbit and in this surface runs the infra-orbital nerve in its bony canal. The floor of the sinus is formed *or by caldwell luc operation*

by the alveolar process of the maxilla and some of the roots of the upper molar teeth may project through the floor into the cavity. The posterior wall, in which runs the posterior superior alveolar branch of the maxillary nerve, forms the front wall of the infra-temporal and pterygopalatine fossae. The medial wall (the base) as seen on the disarticulated maxilla, does not extend much above the floor, but in the living subject the large opening is considerably reduced by encroachment of neighbouring bones, especially the inferior concha. The anterolateral wall forms part of the facial skeleton. The sinus opens on to the middle meatus by means of a small aperture on its medial wall 1 - 2 cm above the level of the floor. The lower end of the frontonasal duct lies very near the maxillary sinus opening. Drainage of the sinus is very poor in the upright position although lying on one side may help to drain one sinus. The nasolacrimal duct is anteromedial to the sinus.

Anteroposterior and lateral X-rays of the skull should be examined in order to confirm the position, size and relations of these sinuses.

The epithelium lining the nasal cavity

The nasal mucous membrane is covered by a respiratory or an olfactory epithelium. The latter is found in the roof and on the neighbouring areas of the medial and lateral walls, and although it cannot be distinguished in the dissected specimen, it is yellowish in the fresh state. The former covers the rest of the cavity except the vestibule which is lined with skin containing vibrissae.

The nerves and vessels of the nasal cavity and paranasal sinuses

The nerves of general sensation are branches of the ophthalmic and maxillary divisions of the trigeminal. The nerve of special sensation is the olfactory. The **olfactory nerves** are the central processes of neurons in the olfactory mucous membrane. Bundles of nerve fibres pass upwards through the cribriform plate and end in the olfactory bulbs of the brain. The arteries largely correspond with the nerves and are branches of the maxillary and ophthalmic arteries. Additional branches enter the vestibule from the upper lip (facial artery). The veins drain into the pterygoid plexus and the facial veins.

FUNCTIONAL ASPECTS

All the paranasal sinuses are small at birth and increase in size during the eruption of the 2nd dentition. Their size is very variable in different individuals. They have the effect of making the head lighter and so reducing the tension in the postvertebral muscles in the upright posture. They are said to add resonance to the voice. In animals they have an olfactory and a temperature regulating function. The sinuses may have persisted in man although they have lost both these functions.

CHAPTER 16

THE JOINTS OF THE CERVICAL VERTEBRAE

INTRODUCTION

IN this last dissection an attempt will be made to examine the joints involved in the movements of the head and the cervical vertebral column. These movements are important because they make it possible for the special sense organs, especially the eyes, to be used effectively.

Revise the attachment of the scalene and other neck muscles (pages 8, 13, 41 and 65).

DISSECTION

Examine the posterior half of the skull and the vertebral column. On the front of the upper thoracic and all the cervical vertebrae are vertical and oblique muscles running between the vertebrae. Where the fibres are confined to the neck the muscle is called the **longus colli** and where the fibres reach the occiput they are called the **longus capitis.** There are also two short muscles between the front of the atlas and the occiput. All the muscles are supplied by the ventral rami of the cervical nerves. When the muscles of both sides contract, they flex the neck and the head.

Remove the muscles from the front of the bodies of the vertebrae and find the strong **anterior longitudinal ligament.** This ligament extends along the whole length of the vertebral column, binding the bodies and discs together. Remove this ligament from two neighbouring lower cervical vertebrae and examine the articulations between the bodies. The major joint consists of the intervening intervertebral disc and a thin layer of hyaline cartilage on the adjacent surfaces of the bodies of the vertebrae. Laterally, on each side of the body, is a small synovial joint between the edges of the bodies of the vertebrae. Remove small pieces of bone laterally in order to open these joints which, though small, are important. They may become diseased and give rise to severe disability and pressure on the spinal nerves.

159

The cervical vertebrae also articulate with each other by means of synovial joints between the articular processes. Examine the skeleton and determine that the plane of the joints between the lower five vertebrae is directed backwards and upwards. The movements at these joints involve gliding and tilting. Upward gliding and forward tilting of an upper vertebra on a lower produces flexion, and the opposite produces extension. An upper vertebra can rotate on a lower about a vertical axis, producing rotatory movement to right and left and can also glide laterally to some extent, producing right and left lateral flexion.

Examine the back of the bodies of the vertebrae by removing the remainder of the posterior arch of the atlas and of the laminae of the cervical vertebrae together with the ligamenta flava. Cut the nerve roots and remove the spinal cord. Find the **posterior longitudinal ligament** running down the backs of the bodies of the vertebrae.

The lateral masses of the atlas articulate above with the occipital condyles at the atlanto-occipital joints and below with the axis at the atlanto-axial joints. The anterior surface of the **dens** of the axis articulates with the back of the anterior arch of the atlas. These are all synovial joints.

Follow the posterior longitudinal ligament upwards. Above the axis it is called the **membrana tectoria** which is continuous with the endocranial dura on the inner aspect of the skull. Cut the membrana tectoria transversely and turn the pieces upwards and downwards. The **cruciate ligament** is now exposed. The vertical part extends from the back of the body of the axis through the foramen magnum to the occipital bone. The transverse part passes between the lateral masses of the atlas behind the dens. This ligament, together with the anterior arch of the atlas, completes the ring enclosing the dens round which the atlas together with the head can rotate.

Cut through the upper part of the cruciate ligament and find the **alar ligaments** which pass from the sides of the dens laterally and upwards to the medial sides of the occipital condyles. They limit rotation. In the midline in front of the upper half of the vertical part of the cruciate ligament is the **apical ligament** which passes from the apex of the dens to the anterior border of the foramen magnum. Open up the atlanto-occipital joints and note the kidney-shaped concavoconvex articular surfaces allowing nodding movements. Open up the atlanto-axial joints and note the round, almost flat articular surfaces allowing rotational movements.

STRUCTURAL DETAILS

The occipital bone (see also page 10)

The occipital bone develops from several centres of ossification and is usually divided into four parts arranged around the **foramen magnum.** The **condylar parts,** situated anterolateral to the foramen magnum have on their inferior surface the kidney-shaped condyles which articulate with the lateral masses of the atlas. The projection lateral to the occipital condyle is known as the **jugular process,** in front of which lies the jugular foramen. Posterior and superior to the foramen magnum, the **squamous part** of the occipital bone extends towards the vault of the skull. The **basilar part** of the occipital bone is thick, extends forwards from the foramen magnum and articulates with the sphenoid bone. In the young skull these two bones are joined by cartilage, but by the twenty-fifth year this has become completely ossified. The part of the occipital bone above the superior nuchal lines develops in membrane but the remainder of the bone forms part of the chondrocranium and develops in cartilage.

The **cervical vertebrae** are described on page 12.

FUNCTIONAL ASPECTS

Balance of the head on the atlas and movements of the head

The head is balanced on the atlas in such a way that without contraction of the muscles, the head falls forwards. Sleeping with the body in the upright position confirms this. Braces to prevent this falling forwards are therefore placed posteriorly and the muscles at the upper end of the vertebral column passing to the occiput act in this way (trapezius, semispinalis capitis, splenius capitis and longissimus capitis). The muscles on the front of the cervical vertebral column are smaller. The large number of small muscles between the occipital bone, atlas and axis enable fine movements to be carried out and permit accurate alignment of the visual axes.

161

Flexion and extension of the head occur at the atlanto-occipital joints. In the upright position, gravity is the main force producing flexion and controlled relaxation of the extensors on both sides of the back of the head regulate the rate and degree of flexion. From the supine position and against resistance, both the sternocleidomastoids and the prevertebral muscles flex the head and the cervical vertebrae. Extension from the prone position is produced by the postvertebral muscles attached to the occiput.

Rotatory movements take place between the atlas and axis, the head moving with the atlas. The oblique muscles of one side of the neck, both the anterior and posterior groups, produce rotation. The right sternocleidomastoid rotates the face to the left; the right splenius capitis rotates the face to the right; the right trapezius (upper fibres) turns the face to the left. Rotation of the head is increased by the rotation of the cervical vertebrae on each other.

Lateral flexion of the head to the left and right takes place at the atlanto-occipital joints and is increased by similar movements between the cervical vertebrae. In these movements the anterior and posterior groups of muscles of one side function together, for example, the right sternocleidomastoid and right trapezius, and the right longus capitis, longus colli and scalenes produce right lateral flexion of the head and neck.

CHAPTER 17

THE LYMPH DRAINAGE OF THE HEAD AND NECK

ALL the tissues of the head and neck, with the exception of the central nervous system, bone and cartilage have a rich plexus of lymph vessels.

The lymph nodes

These are arranged in two groups, the **cervical,** which are divided into superficial and deep, and the **circular,** placed at the junction of the head with the neck. The **deep cervical nodes** form a vertical series extending from the base of the skull to the root of the neck. The majority of the nodes are placed on the outer surface of the carotid sheath deep to the sternocleidomastoid. Some nodes are named. The **jugulodigastric** lies near the angle of the mandible and receives lymph from the tonsil, tongue and teeth. It is frequently palpable. The **jugulo-omohyoid** is situated near the intermediate tendon of the omohyoid. It is important because it receives lymph from the tongue. The **superficial cervical nodes** include the **tracheal** which lie along the inferior thyroid vessels and are continuous with a group of nodes placed behind the manubrium, and the **external jugular** which lie along the external jugular vein superficial to the deep cervical fascia.

The deep cervical group receives the lymph vessels from all the other lymph nodes in the head and neck. In addition direct afferents reach them from the thyroid and salivary glands, the tongue, tonsil, nose, pharynx and larynx. Their efferents unite to form the jugular lymph trunk. On the left side this joins the thoracic duct and on the right it may join the right lymph duct or enter the junction of the internal jugular and subclavian veins independently.

The **circular nodes** are arranged in the following groups:

(*a*) **occipital,** along the superior nuchal line and draining the back of the scalp,

(*b*) **retro-auricular,** on the mastoid process, draining the side of the scalp and the back of the auricle and external acoustic meatus,

(c) **superficial parotid,** immediately in front of the tragus, draining the side of the scalp, and the front of the auricle and external acoustic meatus,

(d) **deep parotid,** deep to the fascia over the parotid gland, with some nodes embedded in it, and afferents draining the parotid gland, the side of the scalp, the lateral halves of the eyelids, the external acoustic meatus, the tympanic cavity, and part of the nasal cavity,

(e) **submandibular,** along the facial artery in the groove between the lower border of the mandible and the submandibular salivary gland with some nodes embedded in its substance and afferents draining the face including the medial halves of the eyelids, the side and vestibule of the nose, the upper lip, the lateral part of the lower lip, the gums, the side of the tongue and the submandibular and sublingual glands,

(f) **submental,** behind the chin in the interval between the anterior bellies of the digastric muscles, draining the central part of the lower lip and the tip of the tongue,

(g) **retropharyngeal,** between the pharynx and the prevertebral fascia draining the back of the pharynx, the auditory tube and the joints of the neck.

Efferent lymph vessels from all the groups of the circular chain drain into the deep cervical nodes.

The tonsils (palatine, nasopharyngeal and lingual) form a ring of lymphoid tissue in the upper part of the respiratory and alimentary tracts in the pharynx and mouth. The efferent vessels from the palatine and lingual tonsils pass to the deep cervical nodes and the efferents from the nasopharyngeal tonsils pass to the retropharyngeal nodes.

Lymph drainage of individual structures

1. **The scalp.** Lymph vessels run with the blood vessels to the superficial and deep parotid, retro-auricular and occipital nodes.

2. **The auricle and external acoustic meatus.** Drainage is to the superficial and deep parotid, and retro-auricular nodes.

3. **The face.** Lymph passes to the parotid and submandibular nodes. Lymph from the central part of the lower lip goes to the submental vessels.

4. **The tongue.** There is a rich plexus of lymph vessels in its submucosa.

(*a*) Those from the tip pierce the mylohyoid muscle and end partly in the submental nodes and partly in the jugulo-omohyoid and adjacent nodes.

(*b*) Those from the margins of the anterior two-thirds pass to the submandibular and deep cervical nodes of their own side.

(*c*) The central part of the anterior two-thirds drains to the submandibular and upper deep cervical nodes on both sides of the neck.

(*d*) The posterior third drains to the upper deep cervical nodes of both sides.

5. **The gums and teeth.** Lymph passes mainly to the submandibular and upper deep cervical nodes.

6. **The tonsil.** Its lymph vessels pierce the superior constrictor muscle and enter the upper deep cervical nodes, especially the jugulodigastric.

7. **The nasal cavity.** (*a*) The anterior part of the nose sends vessels to the submandibular nodes. (*b*) The nasopharynx and posterior part of the nose drain to the retropharyngeal and upper deep cervical nodes.

8. **The tympanic cavity.** Its lymph goes to the parotid and upper deep cervical nodes.

9. **The thyroid gland.** The lymph vessels follow the veins. Those from the upper part drain to the deep cervical group and those from the lower part pass to the tracheal nodes, but some pass along the inferior thyroid veins to the superior mediastinal nodes.

10. **The larynx.** Vessels from the mucous membrane above the vocal folds pierce the thyrohyoid membrane and enter the upper deep cervical nodes. Those from below the vocal folds pierce the conus elasticus and pass to the tracheal and lower deep cervical nodes.

CHAPTER 18

THE SENSORY NERVE SUPPLY

THE skin of the head and neck is innervated by the trigeminal cranial nerves, by branches of the ventral rami of the 2nd, 3rd and 4th cervical spinal nerves and by branches of the dorsal rami of the 2nd to the 5th cervical spinal nerves. The areas of supply by the individual nerves are indicated in Figure 59. The trigeminal nerve supplies the skin in front of a line from the vertex of the skull to the upper edge of the auricle, along the upper half of its posterior edge, forwards above the meatus and the tragus, across the masseter and finally downwards and forwards to the chin. Behind and below this line, the skin is supplied by the cervical spinal nerves (Fig. 59).

The front of the scalp and the upper eyelid are supplied by the supratrochlear, supra-orbital and lacrimal branches of the ophthalmic nerve, the region of the medial angle of the eye and the root of the nose by its infratrochlear branch, and the sides and tip of the nose by its external nasal branch.

The front of the temporal region is supplied by the zygomatico-temporal branch of the maxillary nerve, the region of the zygoma by its zygomaticofacial branch, and the region of the maxilla, the lower eyelid and the upper lip by its infra-orbital branch.

The back of the temporal region and the front of the upper half of the auricle are supplied by the auriculotemporal branch of the mandibular nerve, the cheek by its buccal branch and the lower lip and chin by its mental branch.

Branches of the dorsal rami supply skin on each side of the midline posteriorly. Named branches are the greater occipital nerve (from the 2nd cervical) and the third occipital nerve (from the 3rd cervical).

Branches of the ventral rami supply the remainder of the skin of the head and neck. The lesser occipital supplies the scalp above and behind the auricle, and the great auricular supplies the skin of the posterior surface and the lower part of the front of the auricle, and over the ramus and angle of the jaw. The transverse cervical nerve passes forwards to the region of the laryngeal prominence and the supraclavicular nerves supply the skin of the side of the lower part of the neck and the chest just distal to the clavicle.

The conjunctiva of the upper eyelid is supplied by the ophthalmic nerve through its supratrochlear, infratrochlear and lacrimal branches and that of the lower eyelid by the maxillary nerve through its infra-orbital branch. The conjunctiva over the eyeball is supplied by the nasociliary branch of the ophthalmic nerve.

The mucous membrane of the nasal cavity is supplied by the maxillary nerve through its nasal, nasopalatine and greater palatine branches, except for the upper, anterior parts of the roof, septum and lateral wall which are supplied by the ophthalmic nerve through its anterior ethmoidal branch. The olfactory mucous membrane in the roof of the nasal cavity is supplied by the olfactory nerve. The mucoperiosteum of the frontal, ethmoidal and sphenoidal sinuses is supplied by the ophthalmic nerve and that of the maxillary sinus by the maxillary nerve.

The mucous membrane of the cheek, lower lip, lower gum and floor of the mouth is supplied by the mandibular nerve through its buccal, inferior alveolar and lingual branches, and that of the upper lip and upper gum by the maxillary nerve through its infra-orbital and superior alveolar branches. The anterior two-thirds of the mucous membrane of the tongue is supplied by the mandibular nerve through its lingual branch (general sensation) and by the facial nerve through its chorda tympani branch (taste). The mucous membrane of the posterior third of the tongue is supplied by the glossopharyngeal nerve (general sensation and taste). The mucous membrane of the palate is supplied by the maxillary nerve through its nasopalatine and greater and lesser palatine branches and the glossopharyngeal nerve through its branch to the pharyngeal plexus.

The mucous membrane of the pharynx is supplied by the glossopharyngeal nerve through its branches to the pharyngeal plexus except for the uppermost part which is supplied by the maxillary nerve through its pharyngeal branch, and the lowermost part which is supplied by the vagus nerve through its recurrent laryngeal branch. The laryngeal mucous membrane above the vocal folds is supplied by the vagus nerve through its internal laryngeal branch,

167

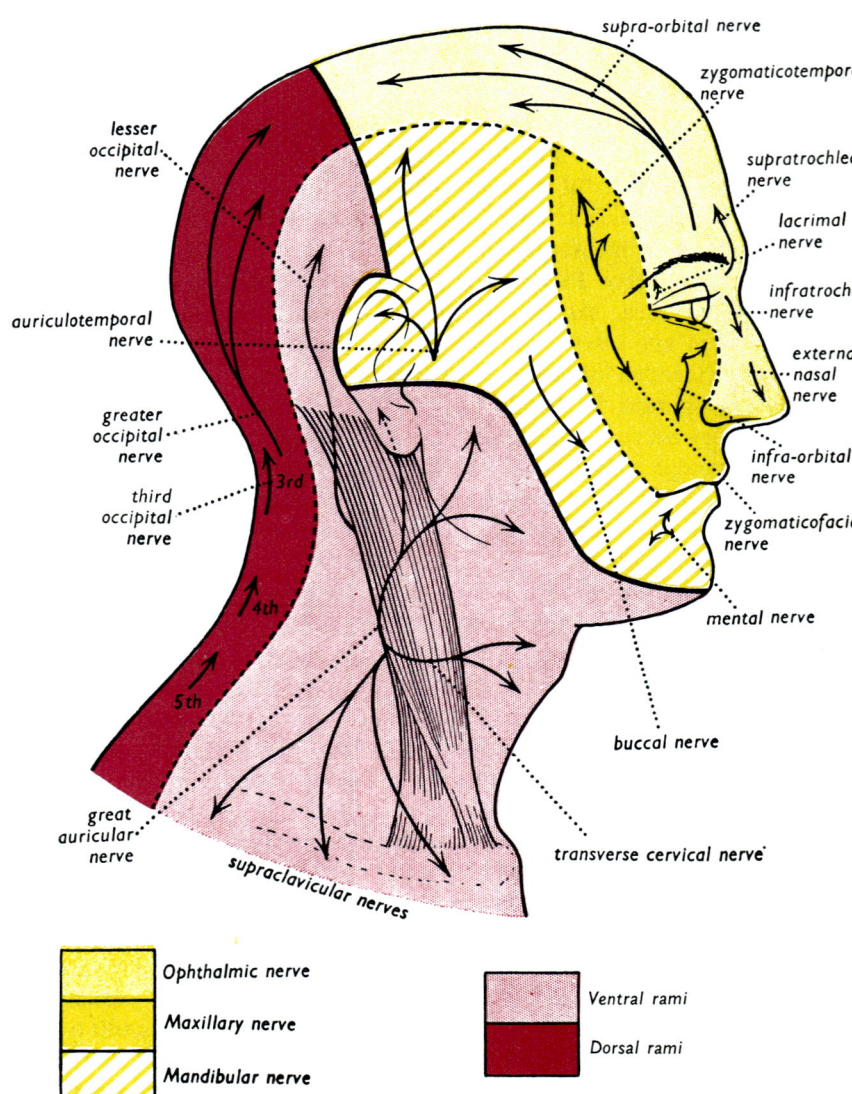

supra-orbital nerve

zygomaticotemporal nerve

supratrochlear nerve

lacrimal nerve

infratrochlear nerve

external nasal nerve

infra-orbital nerve

zygomaticofacial nerve

mental nerve

buccal nerve

transverse cervical nerve

lesser occipital nerve

auriculotemporal nerve

greater occipital nerve

third occipital nerve

3rd

4th

5th

great auricular nerve

supraclavicular nerves

Ophthalmic nerve

Maxillary nerve

Mandibular nerve

Ventral rami

Dorsal rami

Fig. 59

Diagram of the origins of the cutaneous nerves of the head and neck.

168

and below the folds by the vagus nerve through its recurrent laryngeal branch.

The posterior part of the skin of the external acoustic meatus and of the outer aspect of the tympanic membrane is supplied by the vagus nerve through its auricular branch and the rest of the skin and tympanic membrane by the mandibular nerve through its auriculotemporal branch. The mucous membrane of the tympanic cavity, inner aspect of the tympanic membrane and adjacent part of the auditory tube is supplied by the glossopharyngeal nerve through its tympanic branch, and the pharyngeal end of the tube is supplied by the glossopharyngeal nerve through its pharyngeal branches.

PRACTICAL CLASS 1

The osteology described in these classes indicates what should be known about the whole skull and individual bones.

THE OSTEOLOGY OF THE NECK AND OCCIPITAL REGION

Requirements : set of cervical vertebrae, articulated skeleton and skull.

1. The occipital bone

On the occipital bone find
1. the foramen magnum,
2. the occipital condyles (note their shape),
3. the external occipital protuberance,
4. the superior and inferior nuchal lines,
5. the basilar part of the occipital bone in front of the foramen magnum.

2. Cervical vertebrae

On a typical cervical vertebra find
1. the body, the pedicle, the lamina, the spinous process,
2. the transverse process with its anterior and posterior tubercles,
3. the foramen in the transverse process (This is characteristic of cervical vertebrae),
4. the articular processes and facets,
5. the vertebral canal.

3. Vertebral column

In an articulated cervical vertebral column,
1. which spine is the most prominent?
2. (a) in what direction do the superior and inferior articular facets face?
   (b) what movements do they allow?
3. which spinous processes are bifid and which vertebra has no spinous process?
4. what curvature is found in the cervical region of the spine? (At what age in life does this curvature develop?)

5. are there synovial joints between the sides of the bodies?

6. which vertebra has the greatest width?

4. The atlas

1. Note the absence of a spinous process and body.

2. Find the anterior arch with an articular facet on its posterior surface, and the posterior arch.

3. What are the main differences in the shape of the superior and inferior articular facets?

4. Find the foramen in the transverse process and the groove on the upper surface of the posterior arch. What structures lie in the foramen and the groove?

5. The axis

On the axis find

1. the dens,

2. the facet on the front of the dens. With what does this articulate?

PRACTICAL CLASS 2

THE OSTEOLOGY OF THE SKULL

Requirements: Skull and mandible.

1. The vault

On the vault of the skull find

1. the frontal bone,
2. the parietal bones,
3. the occipital bone,
4. the sagittal (interparietal) suture,
5. the coronal (frontoparietal) suture,
6. the lambdoid (occipitoparietal) suture.

2. The lateral aspect

On the lateral aspect of the skull find

1. the temporal bone,
2. the external acoustic meatus,
3. the zygomatic arch,
4. the zygomatic bone,
5. the greater wing of the sphenoid,
6. the temporal line.

3. The front

On the front of the skull find
1. the maxillae,
2. the nasal bones,

4. The parietal bone

On the parietal bone find
1. the sagittal suture,
2. the parietal eminence,
3. the emissary foramen.

5. The temporal bone

On the temporal bone find
1. the mastoid process,
2. the mastoid notch,
3. the tympanic part (forming the major part of the external acoustic meatus),
4. the squamous part,
5. the squamotympanic fissure,
6. the mandibular fossa,
7. the articular tubercle,
8. the zygomatic process,
9. the petrous part,
10. the carotid canal,
11. the styloid process,
12. the stylomastoid foramen,
13. the suprameatal triangle.

6. The frontal bone

On the frontal bone find
1. the supra-orbital margins,
2. the supra-orbital foramina or notches,
3. the frontal eminences.

7. The zygomatic bone

On the zygomatic bone find
1. the temporal process,
2. the frontal process with a flange of bone separating the orbital cavity from the temporal fossa.

8. The maxilla

On the maxilla find
1. the alveolar, palatine, frontal and zygomatic processes,
2. the infra-orbital foramen,
3. the nasolacrimal groove.

9. The nasal cavity

Look into the nasal cavity and confirm that its upper part separates the two orbital cavities. If the lateral wall of the nasal cavity is broken, examine the interior of the maxillary sinus.

10. The mandible

On the outer surface of the mandible find

1. the body,
2. the ramus,
3. the angle,
4. the condylar process with its head and neck,
5. the coronoid process,
6. the mandibular notch,
7. the mental foramen,
8. the alveolar arch with the teeth.

PRACTICAL CLASS 3

THE ANATOMY OF THE HEAD AND NECK IN THE LIVING SUBJECT

1. Identify the following :
   1. the jugular notch,
   2. the clavicle,
   3. the supraclavicular and suprasternal fossae,
   4. the spinous processes of C6, C7, T1 and T2 (Can you feel the spines of C2-C5?),
   5. the transverse process of the atlas between the angle of the mandible and the mastoid process,
   6. the body and greater horns of the hyoid, and note their relation to the lower border of the mandible,
   7. the thyroid cartilage—its anterior border, upper borders and superior horns (Why is the lower part of the cartilage difficult to palpate?),
   8. the cricoid cartilage,
   9. the tracheal rings.

2. 1. What are the vertebral levels of the hyoid bone and the cricoid cartilage?

2. What happens to the laryngeal cartilages in swallowing?

3. Identify and show the actions of the following muscles:
   1. the sternocleidomastoid,
   2. the trapezius,
   3. the platysma.

4. Identify the following:
   1. the external occipital protuberance,
   2. the superior nuchal line,
   3. the mastoid process,
   4. the tragus,
   5. the zygomatic arch,
   6. the condyle of the mandible with the mouth open and shut,
   7. the pre-auricular point on the posterior end of the zygomatic arch immediately in front of the ear,
   8. the frontozygomatic suture,
   9. the orbital margin,
   10. the supra-orbital notch,
   11. the superciliary arches,
   12. the most prominent part of the frontal bone between these arches,
   13. the frontonasal articulation,
   14. the parietal eminence,
   15. the frontal eminence,
   16. the vertex of the head,
   17. the position of the pterion on the side of the head.

PRACTICAL CLASS 4

THE OSTEOLOGY OF THE SKULL

Requirements : Skull with mandible.

1. On the **base of the skull** find the following:
   1. the basilar part of the occipital bone and its pharyngeal tubercle, and the body of the sphenoid bone,

2. the horizontal processes of the palatine bones, and the vertical plate of the vomer,

3. the carotid canal (What direction does the canal take inside the bone?),

4. the jugular foramen,

5. the styloid process and stylomastoid foramen (What emerges from the foramen? What structures are attached to the process?),

6. the mastoid process and the mastoid notch. (Indicate where the sternocleidomastoid muscle and the posterior belly of the digastric are attached.),

7. the mandibular fossa.

2. The mandible

(a) On the inner surface find

1. the lingula,
2. the mandibular foramen,
3. the mylohyoid line and groove,
4. the mental spines (genial tubercles),
5. the area for the submandibular gland,
6. the area for the sublingual gland.

(b) Find the areas of attachment of the following muscles and ligaments:

1. the temporalis,
2. the masseter,
3. the lateral and medial pterygoids,
4. the mylohyoid,
5. the digastric (anterior belly),
6. the genioglossus and geniohyoid,
7. the superior constrictor and buccinator,
8. the sphenomandibular ligament,
9. the pterygomandibular raphe.

(c) Which parts of the bone are covered by mucous membrane?

(d) What is the relationship of the mental foramen to the alveolar edge and to the teeth? How does this vary with age?

3. The pterygoid region of the skull

(a) Find

1. the infratemporal crest,
2. the lateral pterygoid plate,

3. the medial pterygoid plate with its hamulus,
4. the posterior surface of the maxilla and the pterygopalatine fossa,
5. the choanae (posterior nasal apertures),
6. the mandibular fossa and the articular tubercle,
7. the petrotympanic fissure,
8. the foramen ovale,
9. the foramen spinosum,
10. the spine of the sphenoid,
11. the external opening of the bony part of the auditory tube.

(b) Define the areas of attachment of the following muscles:
1. the lateral pterygoid,
2. the medial pterygoid,
3. the temporalis,
4. the masseter.

(c) Through which foramina or fissures do the following pass:
1. the mandibular nerve,
2. the middle meningeal vessels,
3. the chorda tympani nerve,
4. the inferior alveolar nerve and vessels,
5. the hypoglossal nerve,
6. the facial nerve,
7. the internal jugular vein,
8. the glossopharyngeal, vagus and spinal accessory nerves,
9. the internal carotid artery?

PRACTICAL CLASS 5

THE ANATOMY OF THE HEAD AND NECK IN THE LIVING SUBJECT

On the skin mark the following:

1. **the common carotid artery:** draw a line from the sterno-clavicular joint to a point 1 cm below the tip of the greater horn of the hyoid bone. What relation does this latter point bear to the anterior border of the sternocleidomastoid and to the thyroid cartilage? Palpate the common carotid artery at the level of the cricoid cartilage. Can you feel any lymph nodes? In order to examine the neck for enlarged lymph nodes it is necessary to stand behind the subject.

2. **the external carotid artery**: draw a line from the bifurcation of the common carotid to a point immediately in front of the tragus of the ear. This line curves behind the angle of the mandible. Can the pulsation of the artery be felt?

3. **the lingual artery**: draw a line from the external carotid at the level of the greater horn of the hyoid forwards for 3 cm along the upper border of the greater horn.

4. **the facial artery**: its entry into the face (and it should be felt pulsating) can be marked at the anterior border of the masseter where the artery turns round the lower border of the mandible. The point is about 4 cm in front of the angle of the jaw. Its course in the face is tortuous, but it usually passes about 1 cm lateral to the angle of the mouth and then runs upwards in the groove between the nose and cheek.

5. **the superficial temporal artery**: from immediately in front of the tragus vertically upwards for 6 cm, where it divides. The pulse can be felt in front of the tragus where the vessel crosses the zygomatic arch.

6. **the subclavian artery**: from the sternoclavicular joint draw a curved line to the middle of the clavicle rising about 2 cm above the clavicle. Compress the 3rd part of the subclavian artery against the 1st rib by pressing downwards behind the midpoint of the clavicle. By what means will you ascertain that your attempt at compression is successful? What structures can be palpated just above the subclavian artery?

7. **the internal jugular vein**: draw a line from the sternoclavicular joint to the lobule of the ear.

8. **the external jugular vein**: this can be seen in almost all cases passing downwards behind the angle of the jaw to about the midpoint of the clavicle. How can this vein be distended?

9. **the spinal accessory nerve**: draw a line from a point halfway between the tip of the mastoid process and the angle of the jaw to a point just above the middle of the posterior border of the sternocleidomastoid and then to a point on the anterior border of the trapezius 6 cm above the clavicle.

10. **the facial nerve:** the point from which the branches radiate is about 1 cm below the tragus.

11. **the nerve foramina on the face:** draw a line from the supra-orbital notch to the interval between the two lower premolar teeth. The infra-orbital nerve lies on this line 1 cm below the orbital margin. The mental nerve lies on this line 1·5 cm above the lower margin of the mandible. If two premolar teeth are not present, use a point lying 2·5 cm from the midline.

12. **the parotid duct:** the middle third of a line joining the notch below the tragus to the midpoint between the ala of the nose and the red margin of the lip. Check this marking by palpating the duct at the anterior border of the masseter muscle.

13. **the middle meningeal artery:** its position varies but it usually enters the skull slightly in front of the tragus, passes anteriorly and laterally and divides about 2 cm above the middle of the zygomatic arch. The anterior branch curves slightly forwards to the pterion and then backwards towards the vertex. The posterior branch passes backwards just above the upper limit of the attachment of the auricle.

14. **the transverse and sigmoid sinuses:** they are represented by a double line about 1 cm wide passing from the external occipital protuberance towards the external acoustic meatus and then downwards to the tip of the mastoid process along the line of attachment of the auricle.

15. **the thyroid gland :** the lateral lobe extends from the middle of the lamina of the thyroid cartilage as far down as the 6th tracheal ring. The isthmus lies at the level of the 2nd, 3rd and 4th rings.

PRACTICAL CLASS 6

THE OSTEOLOGY OF THE INTERIOR OF THE SKULL

1. On the inner surface of the **vault of the skull** find
   1. the midline groove for the superior sagittal sinus,

178

2. the depressions for the lacunae containing the arachnoid granulations at the sides of the groove,
3. the grooves for the middle meningeal vessels.

2. On the inner surface of the **base of the skull** find the following:
   1. the anterior, middle and posterior cranial fossae,
   2. the orbital plates of the frontal bone separating the anterior cranial fossa and the orbits,
   3. the cribriform plate,
   4. the crista galli,
   5. the lesser wing of the sphenoid,
   6. the anterior clinoid process,
   7. the tuberculum sellae,
   8. the hypophyseal fossa,
   9. the posterior clinoid process,
   10. the dorsum sellae,
   11. the carotid groove,
   12. the superior orbital fissure,
   13. the optic canal,
   14. the greater wing of the sphenoid,
   15. the foramen rotundum,
   16. the foramen ovale,
   17. the foramen spinosum,
   18. the foramen lacerum,
   19. the superior border of the petrous temporal bone,
   20. the fossa for the trigeminal ganglion,
   21. the internal acoustic meatus,
   22. the jugular foramen,
   23. the groove for the transverse and sigmoid sinuses,
   24. the foramen magnum,
   25. the hypoglossal canal,
   26. the condylar canal.

3. What are attached to
   1. the anterior and posterior clinoid processes,
   2. the superior border of the petrous temporal bone,
   3. the margins of the transverse sinus,
   4. the crista galli and edges of the groove in the midsagittal plane of the vault?

4. What passes through
   1. the cribriform plate,
   2. the superior orbital fissure,
   3. the optic canal,
   4. the foramen rotundum,
   5. the foramen ovale,
   6. the foramen spinosum,
   7. the jugular foramen,
   8. the hypoglossal canal,
   9. the condylar canal,
   10. the internal acoustic meatus,
   11. the foramen magnum,
   12. the carotid canal?

PRACTICAL CLASS 7

THE ANATOMY OF THE HEAD AND NECK IN THE LIVING SUBJECT

1. Inspect the **eyes and eyelids** and identify

   1. the palpebral fissure,
   2. the puncta lacrimalia,
   3. the lacrimal caruncle,
   4. the plica semilunaris,
   5. the openings of the tarsal glands,
   6. the conjunctival fornices.

2. With an ophthalmoscope inspect the **interior of the eyeball.** Note the colour of the retina and the arteries and veins converging towards the optic disc which is relatively pale.

3. With an auroscope examine the **external acoustic meatus** noting any changes in direction. To see the drum, pull the auricle gently backwards and upwards. Note the shape, colour and plane of the drum. Running from the centre of the drum upwards and forwards is the shadow of the handle of the malleus. The upper part of the drum above the handle is called the flaccid part. Running downwards and forwards from the middle of the drum is a reflexion of the light of the auroscope called " the cone of light."

4. 1. Examine the **mucous membrane of the mouth and oropharynx.** Note the modifications over the tongue and tonsils. Identify the frenulum of the tongue, the sublingual fold, the opening of the submandibular duct, the lingual artery and veins, the arches of the fauces, the soft palate and the uvula and the opening of the parotid duct.

   2. **Lingual and inferior alveolar nerves;** although there is little point in marking these nerves on the skin, the following relations should be known. The lingual nerve lies deep to the mucous membrane of the mouth on the inner surface of the mandible just below the alveolar margin at the level of the third lower molar tooth. The inferior alveolar nerve enters the inferior alveolar foramen which lies halfway between the anterior and posterior borders of the ramus of the mandible at the level of the crowns of the lower molar teeth.

3. In an individual whose mucous membrane is not particularly sensitive, it should be possible with the help of a mirror to examine the upper part of the **larynx** with its epiglottis, arytenoid cartilages and vestibular and vocal folds. Note the change in position of the vocal folds when a voiced sound is produced.

INDEX

HEAD AND NECK

183

191